The Global Commons

The Global Commons

Environmental and Technological Governance

2nd Edition

JOHN VOGLER

Liverpool John Moores University, UK

JOHN WILEY & SONS, LTD

Chichester • New York • Weinheim • Brisbane • Singapore • Toronto

Other Wiley Editorial Offices

John Wiley & Sons, Inc., 605 Third Avenue,
New York, NY 10158-0012, USA

WILEY-VCH Verlag GmbH, Pappelallee 3,
D-69469 Weinheim, Germany

Jacaranda Wiley Ltd, 33 Park Road, Milton,
Queensland 4064, Australia

John Wiley & Sons (Asia) Pte Ltd, 2 Clementi Loop #02-01,
Jin Xing Distripark, Singapore 129809

John Wiley & Sons (Canada) Ltd, 22 Worcester Road,
Rexdale, Ontario M9W 1L1, Canada

Library of Congress Cataloging-in-Publication Data

Vogler, John.
    The global commons : environmental and technological governance / John Vogler.—
2nd ed.
        p. cm.
    Includes bibliographical references and index.
    ISBN 0-471-98826-X — ISBN 0-471-98574-0 (pbk.)
    1. Environmental law, International. 2. Commons. 3. International cooperation.
I. Title.
K3585.4 . V64 2000
341.7'62—dc21   99-087057

British Library Cataloguing in Publication Data

A catalogue record for this book is available from the British Library

ISBN 0 471 98826 X (hardback) 0 471 98574 0 (paperback)

To
Eirwen, Elen and Thomas

Contents

Preface to the First Edition   xi

Preface to the Second Edition   xv

International Agreements on the Global Commons   xvii

Abbreviations and Acronyms   xxi

1   **The Governance of the Commons**   1
   The Nature of the Commons   2
   The Global Commons   6
   The 'Tragedy of the Commons'   10
   Governance and Regimes   15
   Notes   18

2   **Regime Analysis**   20
   Issue Areas   23
   Actors   25
   Principles and Norms   29
   Decision-making Procedures   34
   Rules   36
   Regime Change   39
   Notes   42

3   **The Oceans**   44
   The Law of the Sea   45
   Whaling   48
   Marine Pollution   55
   The Deep Seabed   61
   Summary   68
   Notes   69

4 Antarctica   73
The Antarctic Treaty System – A Single Regime   77
Principles and Norms   78
Organization and Procedures   83
Rules   85
Monitoring and Enforcement   89
Scientific Activity   90
Summary   91
Notes   93

5 Outer Space   95
The Space Commons and Space Law   99
Military Uses   102
Environment and Space Debris   104
Information Flow   108
Orbit and Spectrum   111
Summary   118
Notes   119

6 The Atmosphere   122
Stratospheric Ozone   124
Climate Change   133
Summary   146
Notes   147

7 Regime Effectiveness   152
Effectiveness as International Law   155
Effectiveness as Transfer of Authority   162
Effectiveness as Behaviour Modification   166
Effectiveness as Problem Solving   175
Regime Assessment   179
Notes   181

8 Explaining Regime Incidence and Change   184
Structural Explanations   186
Utilitarian Explanations   195
Plural Interests and Values   199
Changing Cognitions – Epistemic Communities   204
A Synthesis?   207
Notes   210

9 Conclusion   212
Global and Local Commons   214

Multi-layer Governance   222
Notes   226

References   227

Index   239

Preface to the First Edition

One of the more sobering aspects of writing books about international relations is that as the manuscript goes through the production process, events outside move on. This book began to take shape in 1992 at the time of the Rio Earth Summit (UNCED). In retrospect, the Summit represented a high watermark of government and public interest in international environmental cooperation in much the same way as did its predecessor conference, held twenty years before in Stockholm. While the manuscript was being completed, the paralysis that had afflicted the new legal rules (discussed in Chapter 3) for the most ancient of the global commons, the high seas and deep seabed, suddenly ended. Having been over twenty years in the making, the United Nations Third Law of the Sea Convention entered into force on 16 November 1994.

As the proofs arrived, the first Conference of the Parties to the Framework Convention on Climate Change was convened in Berlin. As argued in Chapter 7, the great project to retard (but not to reverse) climate change was only begun by the Convention signed in Rio in 1992. The Berlin meeting did not produce any major advance on the arrangements outlined in the book; no new specific targets were agreed for the reduction of greenhouse gas emissions or their removal by sinks, and the Parties failed to agree their rules of procedure for future meetings. They did, however, give themselves the 'Berlin Mandate' involving a commitment to negotiating new targets with specified timeframes before 1998. It is to be hoped that the process of responding to the ever more convincing evidence of anthropogenically produced climate change will be effectively established, although the complex politics and economics of regime creation provide many grounds for doubt and disillusionment.

In the early months of 1995 commons issues again provided headline news in the Canadian–EU fisheries dispute. This dramatised the conflict potential of the collapse of fish stocks and associated livelihoods in the over-exploitation of a common resource. The problem, which is briefly touched upon in Chapter 3, is a standard instance of the 'tragedy of the commons'. The North Atlantic Fisheries regime is relatively weak and it remains to be seen whether the current

United Nations negotiations on 'straddling (fish) stocks' can improve matters. There is, thus, no shortage of evidence of the failures of international cooperation in the face of mounting pressure on common resources, which lie outside national jurisdiction, but which are vital to future human welfare. Nonetheless, it is a primary purpose of this book to demonstrate the ways in which cooperation can occur and that it is, despite all the manifest hypocrisy and tardiness of international responses to environmental degradation, indispensable.

The genesis of this particular book can be traced to the early 1980s and the present author's interest in the politics of the outer space commons. This eventually led to the writing of part of the Open University's D312 Global Politics course. I am grateful to Tony McGrew and Paul Lewis at Hall for encouraging me, at that time, to broaden the scope of what started as a regime analysis of international cooperation in space. Since 1990 I have had the good fortune to be involved with ESRC's Global Environmental Change Programme and the British International Studies Association (BISA) Environmental and International Relations Working Group, which it supports. Both have contributed a great deal to my own education and to the writing of this book. The BISA group, in particular, provides an often disputatious but generally congenial setting for the discussion of the international relations of environmental change.

The School of Social Science at Liverpool John Moores University gave me staff development funding which yielded extensive relief from teaching in the Spring Term (we still had them then) of 1993. I hope that the committee members concerned think that there has been an adequate return on their investment. During this period it was important for me (if not perhaps for the students) that I continued to take my Third Year Global Cooperation seminar. It is a well-worn cliché that students make significant contributions to the books of their tutors, but it is nonetheless true. I can only thank my seminar groups for their interest and above all for their forbearance. What is less often mentioned is that academic administration now probably constitutes as great an impediment to research and writing as any teaching responsibilities. In my case, I am grateful to my colleague Peter Saunders, who shouldered what were really my duties, at a critical time. Also at the University, Phil Cubbin demonstrated his mastery of graphics packages and plotters in providing figures in this book. In the Humanities Library Bob Caley and Jim Ainsworth were, as ever, unfailingly patient and helpful.

Several people have contributed, directly or indirectly, to the academic substance of this book. Over the years C. Vishnu Mohan has shared his enthusiasm for and knowledge of radio frequencies, orbit and the ITU with me. I have also benefited from collaboration with Mark Imber who manages to combine unrivalled knowledge of the environmental activities of the UN system with good humour. My colleague Charlotte Bretherton was, throughout, an incisive yet encouraging critic. She also spared time from her own writing to read my drafts with an unerring eye for the pompous, the obscure and the unpunctuated. Roderick Ogley, David Scrivener and Owen Greene all generously read and provided expert commentary on parts of the manuscript. I am extremely grateful to them for giving me the benefit of their specialist knowledge on the deep seabed, polar regions and implementation issues. They most definitely improved the finished book but they are not, of course, responsible for its shortcomings.

Liverpool, May 1995

Preface to the Second Edition

I remember one day, as a second-year student, going to visit the professor on some errand or other. He looked up from a paper-cluttered desk and said, 'Never revise your books Vogler'. While somewhat perplexed at the time, during the past year I have discovered what he meant.

The process of producing a second edition of the *Global Commons* has been quite demanding. To make a book which was first written in the immediate aftermath of the 1992 Earth Summit relevant for the first decade of the twenty-first century, has involved keeping track of extensive developments in international environmental negotiations. This is particularly so in the area of climate change and is reflected in the comprehensive revision of the latter part of Chapter 6. Similar revisions proved necessary in relation to Antarctica, with the entry into force of the Madrid Protocol on environmental protection; in the governance of the seabed, with the establishment of the long-awaited mechanisms of the Seabed Authority; and in various other areas such as the international management of the space debris problem. My overall impression is that despite the scale of the problems and the accelerating pace of technological change, there are some grounds for optimism in the capacity for institutional innovation evident in environmental regimes.

The second edition has also had to engage with a great deal of new academic research and writing on international environmental politics. This has filled some of the gaps in the regime literature that were discussed in the first edition. In particular, there has been an explosion of work on 'global civil society' which has attempted to investigate the non-state dimension of international cooperation. Sustained policy-relevant research into regime effectiveness and implementation has also occurred, and this is reflected in the revisions to Chapter 7. The conclusion to the book, Chapter 9, is entirely new and reflects the development of my own thinking on the relationship between commons regimes at various scales and the question of the role of international cooperation, something that can no longer be simply assumed.

One development has occurred in the five years between the first and second editions of this book which has revolutionized the research process.

This is, of course, the widespread use of the Internet. Whereas there used to be difficulties in obtaining basic documents on aspects of the global commons, there is now a superabundance of information available to anyone with a computer and a modem. On occasion I have referenced web sites in the text, but the reader will be able, as I did, to follow the latest developments by searching the 'world-wide web'.

Finally, I would like to express my gratitude to three individuals who helped me with the second edition by answering my queries and allowing me to benefit from their expertise. They are Klaus Dodds, who specializes in Antarctica, David Egan, who has been researching maritime pollution regimes, and Richard Tremayne Smith at the British National Space Centre. They improved the second edition, but I alone am responsible for its errors and omissions.

Liverpool, January 2000

International Agreements on the Global Commons

General and Environmental

1972 Declaration of the United Nations Conference on the Human Environment (UNCHE), Stockholm (*ILM*, 22: 455; *UN Yearbook*, 1972, 319–321).

1992 The Rio Declaration on Environment and Development. Agenda 21, United Nations Conference on Environment and Development (UNCED), Rio de Janeiro (Quarrie, J. (ed.), 1992, *Earth Summit 1992*, London, Regency Press, or, Robinson, N.A. (ed.), 1993, *Agenda 21: Earth's Action Plan*, New York, Oceana).

The Oceans

1946 International Convention for the Regulation of Whaling (ICRW), Washington (*SMT*, 1: 67).

1972 Convention on the Prevention of Marine Pollution by Dumping of Wastes and Other Matter, London (*ILM*, 11: 1294; *SMT*, 1: 283).

1973 International Convention for the Prevention of Pollution from Ships (MARPOL), London, and Protocol 1978 (*ILM*, 12: 1319 and 17: 546; *SMT*, 1: 320).

1982 United Nations Convention on the Law of the Sea (Third LoS Convention), Montego Bay. Part XI relates to the deep seabed (*ILM*, 21: 1261; *SMT*, 2: 165).

Antarctica

1959 The Antarctic Treaty, Washington (*SMT*, 1: 150).

1972   Convention for the Protection of Antarctic Seals (CCAS), London (*ILM*, 11: 251; *SMT*, 1: 272).

1982   Convention on the Conservation of Antarctic Marine Living Resources (CCAMLR), Canberra (*ILM*, 19: 841; *SMT*, 2: 86).

1988   Convention on the Regulation of Antarctic Mineral Resource Activities (CRAMRA), Wellington (*ILM*, 27: 868; *SMT*, 2: 415).

1991   Protocol on Environmental Protection to the Antarctic Treaty (PREP), Madrid (*ILM*, 30: 1461).

Outer Space

1967   Treaty on Principles Governing the Activities of States in the Exploration and Use of Outer Space Including the Moon and Other Celestial Bodies (Outer Space Treaty), Washington, London and Moscow (*ILM*, 6: 386; UK Cmnd., 3519).

1979   Agreement Governing the Activities of States on the Moon and Other Celestial Bodies, New York (*ILM*, 18: 1434).

1988   Appendix 30B to The Final Acts Adopted by the Second Session of the WARC on the Use of GSO and the Planning of Space Services Utilizing it (ORB-88) (ITU, Geneva, 1988).

1989   Constitution and Convention of the International Telecommunication Union, Nice (ITU, Geneva, 1990).

1992   Constitution and Convention of the International Telecommunication Union, Geneva (ITU, Geneva, 1993).

The Atmosphere

1985   Convention for the Protection of the Ozone Layer, Vienna (*ILM*, 26: 1529; *SMT*, 2: 301).

1987   Protocol (to the 1985 Vienna Convention) on Substances that Deplete the Ozone Layer, Montreal (*ILM*, 26: 1550; *SMT*, 2: 309).

1990   Revisions to the Montreal Protocol, London (*SMT*, 2: 316).

1992   United Nations Framework Convention on Climate Change (FCCC), Rio de Janeiro (UN General Assembly, A/AC.237/18 (Part II)/Add.1, 15 May 1992).

1997   Protocol to the United Nations Framework Convention on Climate Change, Kyoto (FCCC/CP/1997/7/Add.1, 18 March 1998).

The main sources for the texts cited above are abbreviated as follows:

ILM   *International Legal Materials.* A serial publication by the American Society of International Law, Washington.

SMT   *Selected Multilateral Treaties in the Field of the Environment,* vol. 1 (1983) Nairobi, UNEP, and vol. 2 (1991) Cambridge, Grotius.

Brief summaries of over 100 international environmental agreements, including most of those noted above, are provided in Sand (1992), while description and analysis of the various Rio (UNCED 1992) agreements can be found in Grubb *et al.* (1993). Internet websites now provide easy access to treaty texts with up to date data on current status in terms of signatures, ratifications and amendments. UNEP's Information Unit for Conventions site and the IMO site are particularly useful in this respect.

Abbreviations and Acronyms

ABM	anti-ballistic missile
AGBM	Ad hoc Group on the Berlin Mandate (FCCC)
AOSIS	Alliance of Small Island States
ASAT	anti-satellite (weapon)
ASOC	Antarctic and Southern Ocean Coalition (of NGOs)
ATCM	Antarctic Treaty Consultative Meeting
ATCP	Antarctic Treaty Consultative Parties
ATS	Antarctic Treaty System
BSB	British Satellite Broadcasting
BSS	Broadcast Satellite Services (ITU)
BWU	blue whale unit
C band	satellite frequency band 6/4 GHz
C^3I	communications, command, control and intelligence
CCAMLR	Convention on the Conservation of Antarctic Marine Living Resources
CCAS	Convention on the Conservation of Antarctic Seals
CCIR	International Consultative Committee for Radio (ITU)
CDM	Clean Development Mechanism (FCCC)
CEE	comprehensive environmental evaluation (Madrid Protocol)
CFC	chlorofluorocarbon
CHM	Common Heritage of Mankind
CITES	Convention on International Trade in Endangered Species of Wild Fauna and Flora
CNN	Cable News Network (satellite TV)
COMNAP	Committee of Managers of National Antarctic Programmes
comsat	communications satellite
CoP	Conference of the Parties
CoP/MoP	Conference of the Parties serving as the Meeting of the Parties to the Kyoto Protocol
COPUOS	Committee on the Peaceful Uses of Outer Space (UN)

CPR	common property resource
CRAMRA	Convention on the Regulation of Antarctic Mineral Resource Activities
DBS	direct broadcasting by satellite
DOEM	Designated Officer for Environmental Matters (UN)
EC	European Community
EEZ	exclusive economic zone
EIF	entry into force
ELCI	Environmental Liaison Centre International (NGOs and UNEP)
ELV	expendable launch vehicle
EOS	Earth Observation System (successor to US Landsat)
ESA	European Space Agency
EU	European Union
FAO	Food and Agriculture Organization
FCCC	Framework Convention on Climate Change (UN)
FSS	Fixed Satellite Services (ITU)
G7	Group of Seven (major industrialized countries)
G77	Group of 77 (LDCs at the UN)
GATT	General Agreement on Tariffs and Trade
GEF	Global Environment Facility of the World Bank
GESAMP	Group of Experts on Scientific Aspects of Marine Pollution
ghg	greenhouse gas
GHz	gigahertz – frequency of 1 billion cycles per second
GSO	geostationary orbit
HCFC	hydrochlorofluorocarbon
HFC	hydrofluorocarbon
IADC	Inter-Agency Orbital Debris Coordination Committee
IAEA	International Atomic Energy Agency
ICBM	intercontinental ballistic missile
ICES	International Council for the Exploration of the Seas
ICI	Imperial Chemical Industries
ICRW	International Convention for the Regulation of Whaling
ICSU	International Council of Scientific Unions
IEE	initial environmental evaluation (Madrid Protocol)
IFRB	International Frequency Regulation Board (ITU)
IGO	intergovernmental organization
IGY	International Geophysical Year
IIC	International Institute for Communications
IMF	International Monetary Fund
IMO	International Maritime Organization
INC	Intergovernmental Negotiating Committee (FCCC)
Inmarsat	International Maritime Satellite Organization

Intelsat	International Telecommunications Satellite Organization
IOC	Intergovernmental Oceanographic Commission (of UNESCO)
IPCC	Intergovernmental Panel on Climate Change
ITU	International Telecommunication Union
IUCN	International Union for the Conservation of Nature
IWC	International Whaling Commission
JI	joint implementation (FCCC)
JUSSCANZ	Japan, US, Switzerland, Canada, Australia and New Zealand
LDC	less developed country
LEO	low earth orbit
LoS	Law of the Sea
LRTAP	Convention on Long-Range Transboundary Air Pollution
Ku band	satellite frequency band – 14/11 GHz
MARPOL	International Convention for the Prevention of Pollution from Ships
MEA	multilateral environmental agreement
MEPC	Maritime Environment Protection Committee
MLF	Multilateral Ozone Fund
MSY	maximum sustainable yield
NAMMCO	North Atlantic Marine Mammals Commission
NASA	National Aeronautics and Space Administration
NGO	non-governmental organization
NIEO	New International Economic Order
NWICO	New World Information and Communications Order
OECD	Organization for Economic Co-operation and Development
OPEC	Organization of Petroleum Exporting Countries
OSR	orbit spectrum resource
PCB	polychlorinated biphenyl
PFCs	perfluorocarbons
PREP	Protocol to the Antarctic Treaty on Environmental Protection (Madrid Protocol)
Prepcom	Preparatory Committee (for UNCED)
QUELRO	quantified emission limitation and reduction objectives (FCCC)
REIO	Regional Economic Integration Organization
RIIA	Royal Institute of International Affairs
RMP	revised management procedure
RPOA	Recognized Private Operating Agency (ITU)
SAR	Second Assessment Report (of IPCC)
SARPs	standards and recommended practices
satcoms	satellite communications
SBI	Subsidiary Body for Implementation (FCCC)
SBSTA	Subsidiary Body for Scientific and Technological Advice (FCCC)

SCAR	Scientific Committee for Antarctic Research
SDI	Strategic Defense Initiative
SF_6	sulphur hexofluoride
SPOT	Système pour l'Observation de la Terre (French reconaissance satellite)
STAR	Satellite TV Asia Region
SWMTEP	System Wide Medium Term Environmental Plan (UN)
TAC	total allowable catch
TDF	trans-border data flow
UN	United Nations
UNCED	United Nations Conference on Environment and Development, Rio, 1992
UNCHE	United Nations Conference on the Human Environment, Stockholm, 1972
UNCLOS	United Nations Conference on the Law of the Sea
UNDP	United Nations Development Programme
UNEP	United Nations Environment Programme
UNESCO	United Nations Educational, Scientific and Cultural Organization
WARC	World Administrative Radio Conference (ITU)
WMO	World Meteorological Organization
WTO	World Trade Organization
WWF	World-wide Fund for Nature/World Wildlife Fund

1

The Governance of the Commons

And certainly it will be more difficult at the international level than at national levels of decision-making. So locked are we within our tribal units, so possessive over national rights, so suspicious of any extension of international authority, that we may fail to sense the need for dedicated and committed action over the whole field of planetary necessities. (Ward & Dubos, 1972: 294)

The integrated and interdependent nature of the new challenges and issues contrasts sharply with the nature of institutions that exist today. These institutions tend to be independent, fragmented and working to relatively narrow mandates with closed decision processes. . . . The real world of interlocked economic and ecological systems will not change; the policies and institutions concerned must. (WCED, 1987: 310)

These two quotations span 15 years of concern over the deterioration of the global commons. The first comes from a preparatory report written for the United Nations Conference on the Human Environment held in Stockholm in 1972. The second is one of the conclusions of a similar but more extensive document prepared for the UN General Assembly in 1987 by an international body of experts, the Brundtland Commission. It coined the term 'sustainable development' and was a significant precursor to the second United Nations environment conference, the Rio 'Earth Summit' of June 1992 (more properly the United Nations Conference on Environment and Development or UNCED). Since 1972 the need for 'dedicated and committed action' to protect humanity's physical environment has been widely sensed. The enormous publicity surrounding the 1992 UNCED bears witness to that. But the 20 years between the conferences also saw a shift of scientific and then public and political awareness away from a simple concern with matters such as transboundary pollution towards an often apocalyptic vision of global environmental change.

The global commons, areas or resources that do not or cannot by their very nature fall under sovereign jurisdiction, occupy a central position in this vision. As defined in this book, they include the oceans and deep seabed, Antarctica, space and the atmosphere. During the 1980s and 1990s the focus of international concern was upon the last-mentioned. Alarm over the evidence of stratospheric ozone depletion was paralleled by a growing, yet controversial, awareness of the possible relationship between anthropogenic greenhouse gas emissions and climate change. There was and still is a widespread perception that at some time in the future this will threaten the bases of human existence. The other global commons not only provide analogies for dealing with the problems of the atmosphere but are in themselves intimately interconnected in terms of physical processes and the conduct of research into global environmental change.

The point, made eloquently by both Ward & Dubos and Brundtland, is that despite an ever more holistic conception of such planetary problems, the political and institutional framework within which they must be addressed remains hopelessly fragmented. This is nowhere more evident than in the preservation of the global commons, which at once may be seen as belonging to everybody and nobody. It is self-evidently in the long-run common interest to nurture the commons and to prevent the human behaviour responsible for their degradation and the unsustainable depletion of their resources. How this may or may not be achieved in the absence of a world government and in the presence of 180 or more sovereign states and countless short-term exploiters of the commons, is the subject of this book.

The Nature of the Commons

A commons is 'a resource to which no single decision-making unit holds exclusive title' (Wijkman, 1982: 512). The term has its origins in medieval times when pasture and woodland were by custom set aside for the joint use of villagers. In England the common lands were transferred to private ownership in various waves of enclosure in the sixteenth and then the eighteenth and nineteenth centuries. Elsewhere in the world a variety of common property regimes have continued to exist, exercising collective stewardship over fisheries, pastures and irrigation systems. These 'village commons' have a small-scale community basis and 'in many parts of the world, rights to common property resources are all that separates the landless and the land-poor from destitution' (World Bank, 1992: 142).[1] There are, obviously, extreme differences of scale between local and global commons. Nonetheless, it will be argued that the similarities are more than merely metaphorical. Similar problems and institutional principles can be found at both levels, along with some of the same arguments about governance. The abundant evidence on the workings of small-scale commons can have relevance to the much less developed study of their global counterparts. Indeed it would be difficult to discuss the latter at all outside this context (the pervasive use of the 'tragedy of the commons' model referred to below provides a salient example).

An analysis of the commons must start from the fundamental point that defines them – they are not private property. However, any particular commons may include a range of different types of property resource. It may, because of their inherent and natural characteristics, be impossible to bring some such 'goods' under private ownership. Examples would be the atmosphere or outer space (but not celestial bodies like the Moon). Yet in most instances what matters is not the inherent characteristics of the good but the property rules that are applied to it. Such rules constitute changeable human institutions and in consequence mobile goods, such as fish, become different types of property depending on their specific location in relation to these institutions. In the case of a salmon its property status depends upon where it has the misfortune to be caught. In a Scottish stream it is private property, in the high seas it is nobody's property (until caught!), but in an inshore communally managed fishery it is a common property resource (Buck, 1989: 127–128).

The goods referred to above are 'common pool resources' (Ostrom, 1990: 30). Fish, pasture, water supplies even radio frequencies may be appropriated by users from a common stock. There is, however, another type of common resource at the centre of environmental concerns – the 'common sink'. The reference here is to the use of seas, watercourses or the atmosphere as waste disposal systems. The problem associated with this type of property is not one of appropriation of a scarce resource but of 'putting something in – sewage, or chemical, radioactive and heat wastes into water; noxious and dangerous fumes into the air' (Hardin, 1968: 21). Although a single commons may contain common pool and common sink resources that are, in practical terms, related (common sink marine pollution will diminish common pool fish stocks), the problem associated with each type of resource is quite distinct. For common sink resources it is 'not to regulate the withdrawal rate of stock but to control the use of the resource for the purposes of disposal' (Weale, 1992: 193).

The classification and analysis of the various types of communal or public goods has been developed by economists interested in the public/private relationship and the circumstances under which the 'normal' free market distributional mechanisms associated with microeconomic behaviour do not apply. Their preferred term for this is 'market failure' and it provides a justification for governmental intervention and indeed part of the economic rationale for the very existence of the state (Brown & Jackson, 1990: 28–29).

From the perspective of orthodox public sector economics, goods can be classified in two essential ways. The first concerns whether they yield 'rival' or 'non-rival' consumption and the second considers whether it is possible to exclude people from their use or enjoyment. The inelegant term 'non-rival' (sometimes also referred to as jointness of supply) essentially means that one person's use of a good does not deprive others. Such goods are rare even in the commons. Once goods, like fish or minerals, become scarce there will be 'rivalry' between consumers. The same point could be made about radio frequencies or satellite orbital positions where there is competition for the best 'slots' and use by one transmitter can certainly interfere and impinge upon others.

Property institutions concentrate on the question of exclusion. The original characteristic of a commons is 'open access'. It is available for use by anybody and its goods are 'non-excludable' and free for the taking. A commonly used legal description is *res nullius* – literally the property of no-one. This does not imply that, as with undiscovered land (*terra nullius*) during the great age of maritime exploration, the right to exclusive use of such resources could be seized. In international law this is avoided under the doctrine of 'common property'. For over 100 years this has been applied to the fish and mammals of the high seas, where 'no single user can have exclusive rights to them, nor the right to prevent others from joining in their exploitation' (Birnie & Boyle, 1992: 117).[2]

Alternatively, commons resources may be both collectively owned and managed by a community. This will make the commonly owned goods 'excludable'. User rights, shares and rents can be specified. Most commons have in, fact, developed in this way towards some form of regulated exclusion. To distinguish such arrangements from the 'non-exclusive' *res nullius* commons it is helpful to restrict the use of the term 'common property resource' (CPR) regime to them. Renewable yet vulnerable CPRs can be extensively regulated to achieve sustainability by controlling access and allocating quotas. Such activity may rest on the folk-wisdom of centuries or the latest ecological science and complex mathematical concepts of maximum sustainable yield (MSY – employed in fisheries management). Significantly, 'failure' in the commons is seen in terms of 'collapse into open access' (World Bank, 1992: 143). It is also possible to develop a CPR such that the rules for its use take into account ideas of common welfare and equitable distribution along with compensation for those who may be excluded from its full enjoyment. The analogous international legal term is 'common heritage of mankind' or *res communis humanitatis*. Since the late 1960s this has been a central concept in considerations of the status and development of the global commons; unfortunately, it has been more discussed than implemented.

Using the two dimensions of rivalness and excludability it is possible to construct a simple typology of commons resources (Figure 1.1). Scarce common pool resources may be either non-excludable – an open access commons – or a CPR which relies on the ability to exclude, restrict access and allocate shares. The assumption is that if resources are under pressure CPR status gives some grounds for hope that they can be sustainably managed. The case that has

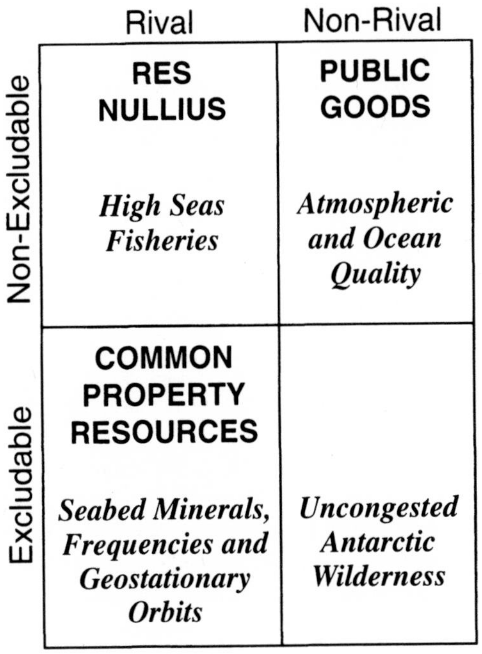

	Rival	Non-Rival
Non-Excludable	**RES NULLIUS** *High Seas Fisheries*	**PUBLIC GOODS** *Atmospheric and Ocean Quality*
Excludable	**COMMON PROPERTY RESOURCES** *Seabed Minerals, Frequencies and Geostationary Orbits*	*Uncongested Antarctic Wilderness*

Figure 1.1   *The status of property in the commons*

excited most theoretical interest is that of public goods, defined in terms of non-rivalness and non-excludability but also non-rejectability. The latter condition stipulates that individuals cannot abstain from consumption even if they might want to (Brown & Jackson, 1990: 34–35). It should be stressed that the pure public good is rare indeed; national defence, street lighting and lighthouses are invariably cited. Beyond this economists tend to differ as to the exact definition of public goods. For our purposes, it is important that environmental economists usually consider that common sinks possess public goods like attributes. More precisely the benefits conferred by common sinks – air quality, unpolluted oceans and a stable climate – can be considered as public goods (Pearce *et al.*, 1989: 12).[3] While there may be disagreement about the exact definition of public goods there is a consensus about the implications in terms of 'market failure'. Public goods cannot be provided by voluntaristic market mechanisms, nobody has an incentive to provide them if there is no possibility of excluding others and charging rent. The state must therefore take on the responsibility for their provision. This conclusion has many implications for the management of commons and most particularly 'common sink' resources.

The final quarter of the quadrant diagram contains those commons resources which are at once non-rival but excludable. An uncongested wilderness like

Antarctica would fall into this category. It is sufficiently vast to ensure that even substantial numbers of tourists, scientists and makers of wildlife documentaries are unlikely to be placed in a situation of rival consumption (but there is a danger that the environment is so fragile that it, and therefore the full enjoyment of the resource by others, could be degraded by quite small numbers of careless visitors). Access is regulated by a select group of Antarctic Treaty States making it an excludable resource. The reference here is to the Antarctic wilderness itself. Antarctic marine resources (krill and cod fisheries) are, as will be seen below, treated as a common property resource which is both in rival consumption and excludable.

The Global Commons

From the seventeenth century the constitutive principle of the modern international system has been the sovereign ownership of territory, coastal waters and more recently airspace. Global commons are areas beyond sovereign state jurisdiction. This may be because of the physical impossibility of extending such control or as a consequence of an international agreement like the 1959 Antarctic Treaty. The limits of the commons and even awareness of their existence have been defined by the current state of exploration and technology. Most of the commons under consideration here could not meaningfully be understood as such before the twentieth century. This was because it was either impossible to reach and exploit them (Antarctica, the deep seabed and space), or they were simply beyond the prevailing state of scientific understanding (the radio frequency spectrum and atmospheric commons).

The oldest recognized commons were the oceans or, more correctly, the high seas. For centuries they were defined as being outside the three mile limit of the territorial sea which came under state jurisdiction. This provides a quaint illustration of the defining role of technology, for three miles approximated to the extreme range of a cannon shot. Beyond this limit the accepted doctrine was that the high seas were open to all for the purposes of navigation, fishing and latterly the laying of cables and aviation. In this period consumption of the common pool resources of the oceans was rival but non-excludable. Once offshore oil exploration and factory fishing became technically feasible, the limits of the high seas commons came under pressure as nations began to press for sovereign economic rights over coastal fisheries and continental shelves. The mining of the deep seabed also became a distinct technical and economic possibility. In a landmark speech delivered to the UN General Assembly in 1967, Arvid Pardo, foreign minister of Malta, suggested that the deep seabed should not be treated as *res nullius* but should instead come under communal international management as the 'common heritage of mankind'. Its supposedly rich mineral resources should be shared amongst the international community on an equitable basis. This proposal, along with pressure for enclosure of substantial areas of the margins of the high seas, made up the substance of the long-running attempt to reform the status of the ocean commons in the Third United Nations Conference on the Law of the Sea (UNCLOS III). The oceans also contain other common property resources: fish and whales. Both were by the 1970s subject to a degree of exploitation that placed the survival of many species in doubt. High seas fisheries retained a '*res nullius*' status alongside regional attempts to create CPRs under the various Fishery Commissions. Whales have, since 1946, been very ineffectually (at least until the 1980s) treated as a global CPR – the focus of much environmental activism and public interest. There has also been increasing alarm at the way in which the oceans were being exploited as a waste disposal system – or common sink. This too has led to attempts at global-scale regulation alongside numerous regional agreements and action plans.

In the early years of the twentieth century the Antarctic represented the last *terra nullius*. During the heroic period of polar exploration a number of states staked formal claims to sovereignty over parts of the polar land mass, but the costs of exploitation and settlement remained prohibitive. Those stations which were set up in Antarctica were dedicated to scientific research, and it was to ensure the continuation of such work and the cooperative relations between scientific personnel, that the Antarctic Treaty of 1959 was signed. This rather exclusive document, originally signed by a group of 12 states, put the various territorial claims into abeyance and outlawed military activity. For the time being Antarctica is by the terms of this agreement a global commons and since 1959 a network of rules has developed for its common use which cover, in particular, the exploitation of maritime resources and the preservation of the unique Antarctic natural environment. The effective collapse of the attempt during the 1980s to provide a set of rules under which minerals might be extracted (as a CPR) has strengthened the international commitment to conservation. Although the signatories to the Treaty will emphasize their role as conservers of the Antarctic for the advancement of science and the wellbeing of the planet, they are very clear that the land mass and its associated marine resources are most definitely not to be regarded as part of the common heritage of mankind.

Contemporaneously with the exploits of Shackleton, Scott and Amundsen, technological changes were opening up another, but rather different, common resource. The invention of radio communications relied upon utilization of a portion of the electromagnetic spectrum (radio frequencies range from 10 kHz to 275 GHz). Frequencies cannot be depleted like minerals or fish stocks or, strictly speaking, be brought under national jurisdiction. They are, nonetheless, a relatively scarce and immensely valuable resource upon which modern civilization routinely depends. Users of the electromagnetic spectrum are interdependent in the sense that some coordination is required in order to avoid mutually damaging interference. Use of the limited frequency spectrum clearly yields 'rivalness in consumption'. Attempts to appropriate frequencies and deny their use to others have implications not only for the efficiency of communications but also for safety. With increasing reliance on mobile maritime and aviation radio an open access radio frequency commons would clearly have had quite literally disastrous consequences. Thus, the problem has been one of the orderly use of the electronic commons and, by implication, methods to exclude and regulate access, something that has preoccupied the International Telecommunication Union (ITU) and its predecessors (the International Radiotelegraph Conferences) since the beginning of the twentieth century.

The exploration of outer space, initiated in 1957, opened up another 'boundless' commons (Schauer, 1977). Within 10 years it was explicitly designated in the UN Outer Space Treaty as the 'province of all mankind' to which there should be free access and which 'is not subject to national appropriation'. The space commons have been exploited in a number of ways: scientific, propagandistic and military. By far the most significant has been the utilization of earth-orbiting satellites for communications, broadcasting and remote sensing. There has, from the beginning, been a close link with the radio spectrum arrangements for the simple reason that the operation and use of spacecraft is reliant on the use of radio frequencies – usually microwave and above. Of paramount importance has been one particular orbit – the geostationary orbit or GSO. At 36 000 km out from the earth's surface this equatorial orbit has the unique property that a spacecraft travelling in the same direction as the earth's rotation will be held stationary in relation to points on the earth's surface. GSO is indispensable to satellite television and to cost-effective telephone and data communication links. It therefore constitutes a global CPR of great value.

In the context of alarm about the prospects for global environmental change – stratospheric ozone depletion and the enhanced greenhouse effect – the atmosphere has come to be regarded as a commons essential to human survival. It may seem inappropriate to characterize the atmosphere in this way. Its constituent gases are not tangible property like seabed minerals or Antarctic wildlife, but are a truly global phenomenon incapable of apportionment under national control. Above all the atmosphere has the properties of a global common sink. The anthropogenic sources of degradation, in chlorofluorocarbons (CFCs), halons and in excessive amounts of carbon dioxide, nitrous oxide and methane are, theoretically at least, under national jurisdiction. The projected effects of continued over-exploitation and consequent degradation have become very well known – even if their degree is disputed. Such consequences are infinitely more serious than a shortage of GSO orbital positions, the inequitable mining of the seabed or exploitation of marine life.

However, in the absence of a world authority the public goods qualities associated with global sinks make them extraordinarily difficult to regulate. 'Perversely, the global challenge is that world cooperation might be least likely in a context where the world stands to lose the most' (Pearce *et al.*, 1989: 12).

'Global' is now such an overworked adjective that some justification is required for its use in connection with commons that have been in the past been described as merely 'international' (Morse, 1977; Caldwell, 1990). A conventional definition of the term 'global' is 'covering, influencing or relating to the whole world'. The spatial scale of the deep oceans and the ubiquitous nature of the atmosphere clearly merit the use of the adjective. GSO is a resource that is global in scope, as are the frequencies and transmissions associated with it, and they are allocated globally (for other radio frequencies this would depend on propagation characteristics, the short range of VHF contrasts with the world-wide reach of HF short wave). Whereas fish stocks are generally dealt with as a local or regional resource, whales, which may transit the oceans, are treated as a global common property resource (the 1946 Whaling Convention has world-wide application).

There is also a deeper meaning to the notion of 'global' commons. It expresses the sense that they are all part of an holistic planetary system and thus interconnected in a range of important and intriguing ways, some of which are only beginning to be perceived. It is appropriate that it was the view from the space commons that gave humankind such a unified image of the earth in all its fragile beauty. In incalculable and often spiritual ways this has affected consciousness of global change, just as the hard science of remote sensing provides much of the evidence.

The global interconnections between the commons and their various resources are extensive. One small example is provided by the link between stratospheric ozone layer depletion and the health of whale populations. Consequent increases in UV/B radiation affect the DNA of the life forms exposed to it and whales swimming in the Southern Ocean are particularly at risk. The most fundamental interconnections are revealed in studies of global climate change. Global circulation models now reflect the beginnings of an understanding of the systematic and complex interdependence between oceans, poles and atmosphere. The role of the oceans in the climate system is acknowledged as a critical area of uncertainty concerning:

> the exchange of energy between the ocean and the atmosphere, between the upper layers of the ocean and the deep ocean and the transport within the ocean, all of which control the rate of global climate change. (Houghton *et al.*, 1990: xxxi)

The stability of the West Antarctic ice sheet (containing ice equivalent to a 5 metre rise in global mean sea level) provides another source of concern in terms of predictions of climate change related sea level rise (Houghton *et al.*, 1990: xxx). The Antarctic is not only an integral part of the global climate system but also provides the indispensable record of past climate change obtained through the drilling of deep ice cores.

The 'Tragedy of the Commons'

Interest in the commons is usually predicated on the belief that they are inherently disaster prone. Overcrowding and technological innovation, allowing ever more intensive exploitation, have led to the depletion and collapse of a range of common pool resources, notably whales and fish stocks. There is acute concern that global 'common sinks' are approaching their carrying capacity. The pervasive metaphor for all this was coined by Garret Hardin (1968) – the 'tragedy of the commons'. It was, in fact, a reworking of an old idea which can be traced back to the writings of the eighteenth century Scots philosopher David Hume (1740). The scenario is a traditional agricultural commons – a pasture open to all where herdsmen may freely graze their cattle. For a time this arrangement may work to the mutual benefit as long as the number of cattle on the pasture does not exceed its carrying capacity. When this point is reached 'the inherent logic of the commons remorselessly generates tragedy'. The incentive for each herdsman is to graze as many cattle as possible for he will receive all the profits from their sale. The negative effects of overgrazing, however, are shared by all the herdsmen and are outweighed by the short-term individual profits from rearing more cattle. In Hardin's words:

> the rational herdsman concludes that the only sensible course for him to pursue is to add another animal to his herd. And another . . . But this is the conclusion reached by each and every rational herdsman sharing a commons. Therein is the tragedy. Each man is locked into a system that compels him to increase his herd without limit – in a world that is limited. Ruin is the destination toward which all men rush, each pursuing his own best interest in a society that believes in the freedom of the commons. Freedom in a commons brings ruin to all. (Hardin, 1968: 20)

Hardin's 'tragedy of the commons' has served as the starting point for most subsequent analyses of commons problems. It falls within what Weale (1992: 205) has called the 'Hobbesian idiom of rational choice theory' and may be regarded as a particular manifestation of the kind of 'social trap' usually modelled by the game of prisoners' dilemma.

Prisoners' dilemma provides a hypothetical demonstration of why rational individuals will fail to cooperate even when it would be in their own joint interest to do so.[4] If, for example, a user of a common pool resource is aware that if he and other users were to exercise restraint the resource would be protected and in fact increased over the longer term to the mutual benefit – it would still be rational for him to ignore this consideration. The basis of this decision would be a lack of trust in other users and a belief that they would exploit his restraint by taking extra shares themselves. This reasoning has obvious relevance to the kind of self-interested distrustful behaviour that Hardin depicts.

The supposed inconsistency between individual and collective interests has also been influentially explored in Olson's (1965) work on the collective action problem. He addresses the problem of the provision of public goods by organizations. Starting from the same premise of 'possessive individualism' as Hardin, the argument is that individuals have no incentive to pay for public goods which they cannot be prevented from enjoying. The rational actor will, therefore, tend to 'free ride'. Coping with 'free riding' will be a problem for any attempt to regulate a common resource, especially if it has the properties of a public good, but, as will be seen, there are various mechanisms and disincentives that can be applied.

These three theoretical models of behaviour all bear upon the condition of interdependence. Their message is that it is very difficult for 'rational' self-interested individuals to manage and resolve problems of interdependence in a cooperative and mutually beneficial way. Interdependence in its simplest form denotes a relationship of mutual vulnerability between actors. It is normal for users of a common pool resource to experience such vulnerability which arises initially from rivalness of consumption. A standard example is provided by the irresponsible use of pastures or fish stocks which will damage the interests of other 'appropriators'. This may also be the case for non-depletable resources such as radio frequencies. Here, users are vulnerable to each other's behaviour in terms of the ways in which they may degrade transmission and reception through signal interference. Vulnerability may extend well beyond the mutual relationship of particular actors involved in direct use of the commons. In the case of common sinks and the destruction or non-provision of public goods, whole populations may be placed at risk. It is useful to distinguish this type of 'common fate' interdependence from more limited and specific forms of 'actor interdependence' (Jones & Willetts, 1984: 32). When considering the projections of global environmental change, related to the overloading of common sinks, we are confronted with the most extreme form of 'common fate' interdependence.

Following the logic of Hardin and other rational choice theorists, actors (who may be well aware of the extent of their long-term interdependence) will still behave towards common resources in a perversely individualistic and short-term way. This diagnosis of the 'tragedy of the commons' leads to two alternative prescriptions for its avoidance. Either common property must be translated into private property, or the state or some other external authority must intervene to enforce good behaviour or provide public goods. The first solution is the one which is presumed to have occurred in the case of the English pastoral commons through the process of enclosure. Individuals can be trusted to husband and use their own land efficiently and sustainably, but not that which belongs to the community or indeed to nobody. Naturally, some common pool resources and sinks cannot be enclosed and here, as in the provision of public goods, the answer lies in action by a coercive state authority. Hardin (1978: 317) is quite clear about the Hobbesian basis of this prescription and actually speaks of the necessity for a Leviathan and 'mutual coercion, mutually agreed upon'.

The reason for this is not only a Hobbesian pessimism or an assumption of possessive individualism, but an argument about the costs and complexity of organizing collective action on a large scale (the economist's term is trans-action costs). That there was an inverse relationship between the likelihood of the voluntary provision of public goods and the size of the group involved was one of Olson's (1965) major conclusions. However, as with the tragedy analogy, Hume had made the essential point two centuries before:

> Two neighbours may agree to drain a meadow, which they possess in common: because 'tis easy for them to know each other's mind; and each must perceive that the immediate consequence of his failing in his part, is the abandoning of the whole project. But 'tis very difficult and indeed impossible, that a thousand persons shou'd agree in any such action: it being so complicated a design, and still more difficult for them to execute it; while each seeks a pretext to free himself of the trouble and expence [sic], and wou'd lay the whole burden on others. (Hume, 1740: 538)

Neither the assumptions nor the solutions associated with the 'tragedy of the commons' model have gone unchallenged. The former may be seen as a displaying a narrow, pessimistic, instrumental and indeed economistic view of human behaviour. It represents one philosophical position which, even its advocates admit, amounts to a necessary over-simplification. The analysis proceeds exclusively from the individual and takes no account of the significance of consciousness of community or collective moral concerns. These may well constitute the 'social glue' that holds a collective enterprise together in ways that are incomprehensible to possessive individualism. Even at the individual level there is neglect of the possibility of altruism and of the significance of gender. Feminists justifiably ask whether herdswomen would behave in the same uncooperative way as Hardin's herdsmen.[5] Finally, the assumptions underlying the 'tragedy' model may be regarded as culture bound, part of an ideology of free enterprise capitalism that has helped to destroy traditional CPR arrangements in the developing world, which had previously been sustained by entirely different and communitarian values (*The Ecologist*, 1993).

The other objections to the model are empirical. It is argued that the English pastoral commons referred to were never open access resources in the first place but a CPR. Villagers were quite capable of regulating the use of common pastures but their institutions were destroyed by the intrusion of new forms of entrepreneurial capitalist agriculture rather than the cupidity of the commoners. Most significantly, there is now an extensive literature describing a rich variety of common property resource regimes to be found throughout the world (Berkes, 1989; Ostrom, 1990). They are under threat, but they can promote sustainable development. The threats to their existence involving a collapse into 'open access' relate to challenges to those communal CPR institutions that according to Hardin and other 'tragedy' theorists ought not to have existed in the first place! The threats are familiar from the English experience, external market changes, population pressure and the introduction of new technologies – in one instance something as simple as nylon fishing nets (World Bank, 1992: 142–143).

At least on a small scale, self-organizing and self-managed common property arrangements can flourish, without the need to rescue them from collapse through privatization or coercive external regulation. Hardin correctly identified the overgrazing problem associated with an open access *res nullius*, but the crucial point is that he does not recognize the possibility of a viable alternative to privatization or coercive management in the form of communal CPR institutions or the collective provision of public goods. The logic of this neglect is impeccable but only under the assumptions of possessive individualism.

A concentration on 'tragedies', the interdependence of commons users and the question of collective action should not be allowed to obscure another perspective on the commons. For many the real tragedy is not that portrayed by Hardin, but the tragedy of dispossession (Middleton *et al.*, 1993: 109). The English enclosure 'solution' resulted in large numbers of dispossessed labourers, 'sturdy beggars' who roamed the countryside. Even before enclosure some cultivators were better placed to profit from common property resources than others. 'No amount of common land was of any use to those peasants who did not own a pig to let loose on it' (Imber, 1988: 154). Where CPR arrangements had developed they were predicated on defined user rights and many of the poor and landless were excluded altogether. Thus the uncritical use of the term 'interdependence' in relation to the commons may be misleading. Describing the relationship between actors as interdependent might imply that they were equal in terms of their share of the benefits derived from common resources and above all in terms of the impact and responsibility for any environmental 'tragedies' that may occur. This is unlikely to be so.

Any commons institutions must be considered within their economic and political context. In the case of the global commons this is an international political and economic system marked by extreme inequalities and in which the relationship between developed market economies and less developed countries (LDCs) is often more one of dominance and dependence than interdependence. A rectification of such inequities, in relation to the benefits that might be drawn from the commons, is at the heart of Common Heritage of Mankind concept (itself developed in tandem with the demand for a New International Economic Order (NIEO)).

Much of the discussion of common pool resources is couched in terms of the efficient use of resources and 'optimal' outcomes.[6] The analysis considers the relative economic efficiency of centrally allocative or market solutions in relation to common or private property regimes (Wijkman, 1982; Denman, 1984). In recent years this has been overlain by explicit economic consideration of the environmental consequences of the 'efficient' exploitation of the commons and above all global common sinks (Pearce *et al.*, 1989; Pearce, 1991). From the perspective of the 'dispossessed' South, such discussions must seem insignificant in relation to the immediate necessities of economic development and physical survival. Southern governments often represent this imperfectly and pursue a variety of other agendas, but they are still unlikely to subscribe to new arrangements for the global commons that do not address the questions of development and equity.

The point is most graphically illustrated in discussions of the future of the atmospheric commons and climate change. For the developed world there are trade-offs between economic growth and environmental quality and issues of intergenerational equity arise. Even in the simple 'tragedy of the commons' model the full impact of decisions to overgraze the pastures may well only be visited on the next generation of herders. The extent to which the possibility of this will be taken into account in current decision-making will depend upon the 'discount rate' that is applied to the future. In relation to the global commons the anticipated consequences of present activities may appear to be highly uncertain and, in any case, several generations distant. The way in which such projections are handled in different societies will be very much a function of affluence or poverty. Members of a society on the margin of starvation will apply a very high discount rate to even the most immediate future.[7] Contemplating the state of the earth's ecosystems in the next century might be seen as a luxury only the rich can afford. Ameliorative measures which require general sacrifices of economic development or growth (for example in the reduction of fossil fuel burning) are hardly likely to be accepted by developing countries when the despoliation of the commons (and the economic benefits arising from it) are so clearly the responsibility of the developed world. Neither is it certain that the consequences will be equally borne despite the metaphors of 'spaceship earth' or ideas of 'common fate' interdependence. In the case of anticipated global warming there are a variety of predictions, concerning the inundation of certain low-lying areas arising from sea level rises associated with the thermal expansion of the oceans. But there are also predictions that a changing climate will simply redistribute benefits: some existing arable land will become arid, while previously infertile areas will flourish. As a generalization it is likely that, given the already marginal existence of many LDCs, the location of major Third World cities at sea level and a dearth of the technical and financial resources necessary to adapt to change, they will be the main sufferers (much the same could be said of the degradation of the ocean commons). In relation to other commons the picture may be rather different. Uncoordinated over-exploitation of the frequency spectrum leading to a reduction in efficiency would, for example, be most damaging to advanced users with enormous investments tied up in electronic communications.

The complex interrelationship between economic efficiency and growth, national interests, scientific uncertainty, environmental survival and equity at various spatial and temporal scales provides the context in which the development of the global commons must be considered. The difficulty in equating all these factors explains why Brundtland's concept of 'sustainable development', ensuring that current activities do not disadvantage future generations, is at once so politically attractive yet difficult to pin down. It is all some way removed from the known propensity of an excessive number of cattle to graze away a piece of medieval common land.

Governance and Regimes

The solution to Hardin's 'tragedy of the commons' is provided by the state. Such is also the conclusion of the argument about public goods. When 'commons' problems of pollution or land use are essentially local in scope they will be within the jurisdiction of a state (unless they have transboundary characteristics) and there is, therefore, a government which at least has the potential to take control and regulate in the collective interest. User rights may be limited, defined and enforced. Taxes on emissions or other damaging activities may be levied, public goods may be provided. This is not to say that such action is always well directed or efficient – there is much evidence to the contrary – but, following Hardin, the first resort of those concerned with the degradation of the commons is most frequently state regulation. The alternative 'tragedy of the commons' solution of enclosure also involves the state. In the case of the English pastoral commons it was state legislation that transferred communal property to private hands and state power that enforced enclosure.

Strictly speaking, such alternatives are unavailable for the global commons. They are global commons because they do not fall under the jurisdiction of any state and the defining characteristic of the international as opposed to domestic systems is the absence of a superordinate authority or world government. Accepting, for the moment, that enclosure might be a desirable solution, it has been shown to be possible through a series of unilateral actions by sovereign states (within a set of common understandings). Substantial parts of the *res nullius* ocean commons were enclosed by the extension of 200 mile Exclusive Economic Zones (EEZs) during the 1970s.[8] Elsewhere it would either be physically impossible (the atmosphere), economically irrational or politically unsupportable. There are good economic arguments for not subdividing a common pool resource into private units when, as in the case of satellite orbits or seabed resources, there is uncertainty as to the definition and value of units and property rights would be difficult to enforce (Wijkman, 1982). Economic rationality cannot be divorced from political realities and although there are territorial claims on Antarctica, attempts at enclosure would generate disproportionate and probably irresolvable political conflict.

The 'state of nature' for the global commons is *res nullius*, but as demonstrated in the development of the Law of the Sea, technological change and pressure on limited common pool resources will exert pressure for collective regulation. This would usually mean a movement from open access to either enclosure or the development of CPR arrangements at the international level. The problem of 'common sinks' in relation to global environmental change is even more acute and most would agree that it cannot be alleviated without moving beyond institutionalized neglect. However, it raises the, for many theorists, insoluble problem of the provision of international public goods. The essential difficulties here relate to the institutional arrangements that define the way in which common pool resources are held and exploited and the way in which common sinks are used. Joint management may be essential to address 'planetary necessities' but, as Brundtland points out, institutions are 'independent and fragmented'. Substantial work has been done on the strictly economic solutions to commons problems, outlining, for example, schemes of international carbon taxation or tradeable permits. There is far less understanding of the basic political and institutional conditions that would enable such arrangements to be put in place (Grubb, 1990; Pearce, 1991).

The problem of the lack of a world authority, and the difficulties of cooperation between sovereign states in its absence, have long been at the heart of the academic study of International Relations. Because there is no natural harmony of interests, behaviour in the global system requires some coordination if mutually damaging outcomes are to be avoided. Discussion was (and continues to be) dominated by the problem of war. Abiding insecurity and the sheer human and material damage inflicted by large-scale hostilities were understandably regarded as the essential problem arising from the ordering of the world into separate sovereign states. The solutions proposed were frequently utopian in their advocacy of world government, but on a more practical level focused on the creation of formal organizations and the strengthening of international law. The emergence of a world economy, increasingly integrated across national frontiers, and an awareness of a wide range of transnational or even global-scale phenomena has served to heighten the contrast between Brundtland's 'real world of interlocked economic and ecological systems' and a political system fragmented into sovereign nation states. In earlier periods solutions might have been advocated that envisaged the creation of some form of world government organization, overriding national sovereignties and having supranational powers. The current use of the term 'governance' is not a synonym for such a government. Rather it implies that there are certain control functions 'that have to be performed in any viable human system' which do not require formal government (Rosenau & Cziempiel, 1992: 3). The concept of governance has been widely used because students of international and domestic political systems have noticed that in both arenas and in the extensive connections between them, formal governmental authority cannot account for all the ways in which human activity is regulated and coordinated (Campbell *et al.*, 1991). Governance is regarded as a 'more encompassing phenomenon than government' that also 'subsumes informal non-governmental mechanisms' (Rosenau & Cziempiel, 1992: 4).

The questions asked in this book concern the forms of governance that actually exist or may be created for the global commons. Since the mid-1970s problems of international cooperation and what is now defined as governance have been studied in terms of the concept of 'regime'. A regime is regarded as an institution or, more precisely, a set of norms, principles, rules and decision-making procedures that govern a particular issue area, such as trade, money or more relevantly the use of the global commons. The concept is explained and developed as an analytical device in Chapter 2. For the present, it is worth stressing that regimes are more than international organizations or even formal legal arrangements between states. They comprise the whole range of understandings, rules and procedures that exist in relation to common pool resources or sinks. Regimes could merely register a minimal rule that a resource is open access (indeed the term *'res nullius regime'* is commonly used). However, they could also be the functional equivalent of government, serving to provide international public goods and to reduce the associated transaction costs (Keohane, 1984). Significantly, for our purposes, international CPR arrangements constitute a relatively developed type of regime serving to coordinate and monitor behaviour.

The focus of much of this book is on analysis of the particular governance functions that regimes can fulfil in relation to the commons. There are evidently enormous differences between small-scale commons institutions and global regimes, but detailed knowledge of the former can at least provide an analytical yardstick for assessing the latter. (The actual and theoretical connections between them will be investigated in the concluding Chapter 9.) The functions that one would expect to find in well developed commons institutions at all levels include the making of collective choices, the setting of standards, rule-making, monitoring of compliance and the generation of knowledge.

In Chapter 2 regime concepts are explained and extended to cover the particular functions of commons institutions. Chapters 3 to 6 attempt to use this analytical framework comparatively in order to establish both the incidence and characteristics of regimes across the various global commons: oceans, Antarctica, space and atmosphere. This provides a basis for an investigation of the meanings of effectiveness in commons governance regimes in Chapter 7. Effectiveness is not easy to define. It has legal, authority and compliance dimensions. Ultimately it requires some judgement of impact upon the fate of the commons, measured in terms of the avoidance of tragedies and sustainability, but also of equity.

The final substantive chapter (Chapter 8) considers issues that have dominated the academic study of regimes – the explanation of change, stability and effectiveness. Regime analysis is essentially a way of laying out the patterns of governance that exist in a system without formal government. There are theories about regimes but no integrated regime theory. What is striking is not that the theoretical approaches to regimes derive from the realist and liberal paradigms that have, for over 50 years, directed the academic study of International Relations, but that there is such a close fit with the debate about the domestic commons. The 'tragedy of the commons' model shares key Hobbesian assumptions with realism and both arrive at the same rather pessimistic conclusion that the degradation of the commons cannot be avoided without enclosure or the intervention of a coercive hegemonic power. From this perspective the role of regime institutions *per se* is likely to be marginal.

On the other side, the liberal institutionalist view of regimes shares with the opponents of the Hardin model a belief in the possibility of self-managed, self-interested, but ultimately enlightened, cooperation between users and beneficiaries of the commons. Effective CPR institutions can be built and these may even encompass the principles of common heritage. Regimes may provide ways of overcoming the transaction cost problem of cooperation within large groups that was first identified by David Hume. There is, thus, a more optimistic answer to the questions about governance posed by Ward & Dubos and Brundtland at the beginning of this chapter.

Notes

1. For an argument that relates global environmental degradation to the enclosure of the very extensive and varied small-scale commons throughout the world see *The Ecologist* (1993). Academic analyses of commons and their governance arrangements are provided by Berkes (1989) and Ostrom (1990).

2. The legal doctrine of common property which applies to *res nullius* open access commons should not be confused with the idea of a common property resource or CPR which, as used here, applies to a good from which users may be excluded and is under some form of communal regulation.

3. The case of the atmosphere is interesting because, as Cliff Baker pointed out to me, increased use as a common sink might be seen to lead to rival consumption – pollution of the atmosphere certainly does degrade the use and enjoyment of others. Logically, then, the atmosphere did have the qualities of a public good in previous centuries when it was nowhere near its carrying capacity and pollution by one individual had no effect on others, but this is no longer the case. The same point could be made about marine pollution.

4. The fundamentals of this analogy are that two criminals are arrested for the same crime and interrogated separately. They are offered lenient treatment if they confess and implicate their accomplice but can achieve the best result if they stay silent, depriving the police of evidence for a conviction. However, staying silent incurs the risk of a prison sentence if the accomplice 'turns Queen's evidence'. Both prisoners would do well to cooperate by staying silent, but as neither has sufficient trust in the other to take this course, the rational strategy, avoiding the risk of the worst alternative, is to confess. This simple game has been very extensively studied and applied to a whole range of 'social traps'. Although it can be used to demonstrate the 'tragedy of the commons' thesis, it is also possible to support the other side of the case through game theoretic experiments of the type carried out by Axelrod (1984). Using an iterated rather than one shot prisoners' dilemma game it can be demonstrated that the 'winning' strategy is to cooperate (and retaliate against opponent's defections). This finding has important implications for the functioning and design of institutions.

5. I am indebted to Charlotte Bretherton for this point and for the observation that in most parts of the developing world it is women who bear responsibility for tending communal resources.

6. The standard concept employed is 'Pareto optimality'. This may be explained by considering a change in the distribution of resources resulting in a situation where some people are better off but without at the same time making anyone worse off. Such a situation is 'Pareto optimal'. This is not the same as fair or equal shares to a common resource and would often be politically unacceptable because it does not deal with the relative deprivation felt by those whose welfare is not improved. Such is the stuff of politics.

7. The reference is more to the social discount rate than the strictly economic version. The idea of social discounting expresses the relationship between the value placed upon present and future consumption. A high rate will encourage immediate consumption, discounting future consumption needs. Individuals and societies on the edge of starvation can 'take little thought for the morrow'. For the strictly economic concept of discounting and its contradictory implications for the environment, see the discussion in Pearce *et al.* (1989: 132–152).

8. This extension, which now has the status of customary international law, was formally negotiated as part of the Third Law of the Sea Convention, which only entered into force at the end of 1994. It is arguable that, in resource conservation terms, it has had some beneficial effects by extending national control over what were common fisheries, inadequately regulated by multilateral Fishery Commissions.

2

Regime Analysis

According to what has become the standard definition, regimes are:

> sets of implicit or explicit principles, norms rules and decision-making procedures around which actors' expectations converge in a given area of international relations. *Principles* are beliefs of fact causation and rectitude. Norms are standards of behaviour defined in terms of rights and obligations. *Rules* are specific prescriptions or proscriptions for action. *Decision-making procedures* are prevailing practices for making and implementing collective choice. (Krasner, 1983: 2, my italics)

They are institutions in the sociological sense of the word: 'social practices consisting of easily recognised roles coupled with clusters of rules or conventions governing relations among the occupants of these roles' (Young, 1989: 32). The study of regimes is not the only approach available. Soroos (1986) has, for example, proposed that the development of international governance can be understood in the ways analogous to the domestic policy process. However, regime thinking has provided the 'dominant paradigm' for American (and inevitably other) discussions of international cooperation since the mid-1970s.

For this reason, if no other, the term 'regime' is unavoidable even if, for the uninitiated, it serves to confuse and misrepresent. In ordinary usage the word denotes a system of government or a particular administration. This is exactly what regime analysts in International Relations and International Political Economy do not mean when they employ the word. Rather, their concern is with forms of governance in the absence of government! Amongst sovereign states the performance of governance functions must involve some element of cooperative behaviour simply because there is no central enforcing power.

The concept of a regime as a social institution is not (as was evident in the previous chapter) exclusive to the study of International Relations. It is used by natural resource economists and ecologists in their discussion of the institutions for the management of common property. Gibbs & Bromley

(1989: 22) refer to common property regimes as 'forms of management grounded in a set of accepted social norms and rules'. Because of this equivalence, regime thinking appears particularly relevant to the study of the governance arrangements for the global commons.

International lawyers have habitually spoken of regimes as a corpus of legal rules relating to a particular subject as in the 'human rights regime' or 'the maritime regime'. This is an overlapping but not coincident use of the term because of the stress that regime analysts have placed upon informal practices and agreements. Ruggie (1975), who is often credited with the introduction of the concept, argued that the existing literature on technological change and international cooperation was inadequate because of its restricted focus on law and organization. What was required was a wider view encompassing implicit understandings between a whole range of actors, which would not necessarily be states. In essence, a regime was an institution that might comprehend some legal rules and some types of formal organization, but went well beyond them. There would, after all, be little point in utilizing the concept if it was merely a synonym for the existing categories of international law and organization. This point is easy to establish in studies, for example of the international monetary regime, where tacit understandings amongst central bankers have played a major role or where a regime is based on long-hallowed practice and usages. In a number of the cases under consideration in this book it is less easy to locate such significant informal components. This is because governance, or perhaps the attempt to provide governance, has frequently involved the active construction of a legal framework rather than the slow accretion of practice. Thus, in their study of global environmental politics, Porter and Brown (1991: 20) find it 'difficult to identify norms or rules in the global environmental area that are not defined by a specific agreement'. Accordingly regimes are defined exclusively in terms of 'norms and rules specified by a multilateral legal instrument among states to regulate national actions on a given issue'. This is too narrowly legalistic and runs the risk of avoiding those critical tacit understandings that may make the bare bones of a legal text both comprehensible and workable.

The impact of institutions as a significant variable in their own right is disputed. Realists and Marxists dismiss regimes as epiphenomena which reflect, but do not substantially affect, the crucial underlying patterns of power or ownership of the means of production.[1] Elsewhere there is a growing realization that institutions constrain and shape economic behaviour by providing, amongst other things, the definitions of property rights without which markets cannot function (Kratochwil, 1989; North, 1990; Young, 1994). This understanding was central to the discussion of common property in the previous chapter. Some would go further than this and argue that regimes can contain norms and principles which are in themselves recognizable as the cause of certain actions.[2] The implication of these differences is that the concept of regime can never be entirely neutral and the 'evident liberal content of much regime theory, particularly in its emphasis on global or collective welfare, indicates the assumptions of a liberal-pluralist framework' (Tooze, 1990: 211–212). It is, thus, as well to be clear at the outset that adoption of a regime approach makes certain assumptions about the role of institutions in determining behaviour, which although widely shared across the social sciences, are not unchallenged.

One danger in theorizing about regimes is that it is all too easy to slip into an obsession with the preservation of order at the expense of questions of justice. The academic study of International Relations inevitably tends to reflect contemporary political problems (as illustrated by the great debates of the 1920s and 1930s over realist or idealist approaches to security). In the case of regime analysis its original popularity in the mid-1970s was a fairly direct response to what appeared to be a recession of American leadership and the dissolution of the post-1945 Bretton Woods arrangements. The problematic was quite self-consciously 'how can order and stability be restored to the international political economy?'. It is, therefore, important to be intellectually self-conscious about exactly what is meant by the success or effectiveness of a particular regime.

By choosing to employ regime analysis the present study necessarily assumes that institutions play a significant role in shaping behaviour. It does not, however, assume that the problematic is the maintenance of order. Otherwise there is an inherent risk that order and international cooperation come to be regarded as desirable, in themselves, without specific reference to any effectiveness criteria. Deciding what constitutes effective governance of the commons is not easy, although an attempt will be made in Chapter 7. The guiding concern in this undertaking will be the extent to which the dilemmas of sustainable yet equitable use of the commons, discussed in the previous chapter, are addressed by the various regimes.

This is not the place to review the voluminous regime literature. Such a task has been ably performed by a number of authors: Haggard & Simmons (1987), Tooze (1990), Milner (1992), Underdal (1995), Stokke (1997). They identify a number of strengths and weaknesses. A few may be singled out. At the outset there was an excessive concentration (understandable in the light of the origins of regime analysis) on a handful of economic regimes. For a considerable period Young (1977, 1982, 1989) was almost alone in applying regime analysis to environmental and natural resource issues, but by the end of the 1990s they had become a major focus of regime thinking. A more fundamental point about regime scholarship in general is that it has tended to neglect analysis and description in favour of large-scale theoretical debates about hegemonic decline and regime creation and change. The various schools of thought will be considered briefly below and in more depth in Chapter 8, but it should be noted that, strictly speaking, they do not constitute a theory of regimes. It is commonplace to speak of 'regime theory' but in effect there is no such thing. Instead there are a number of 'theories about regimes', most of which have a clear ancestry in the classic theoretical approaches of realism, liberal-idealism or functionalism (and an evident relationship to parallel debates on local commons). Much of this literature has, unlike its local counterpart, lacked a substantial empirical base. The study of international environmental regimes has begun to rectify this with substantial studies of the actual processes and mechanisms of governance (Stokke, 1997). Above all, there has been a turn away from the long-standing debates on regime causation and maintenance towards a concern with the consequences and outcomes of international institutions and the problems of evaluating effectiveness (Underdal, 1994; Haas *et al.*, 1993; Victor *et al.*, 1998).

It may be argued that in the absence of a set of validated propositions about regimes the most useful attribute of regime thinking rests with its potential as a tool of comparative analysis. By this is meant the utilization of regime concepts to break down, describe and compare the component parts of these institutions. Surprisingly little such work has been performed. Zacher with Sutton (1996) provides an example of an equivalent enterprise to that attempted in this book with its comparative treatment of international transport and communication regimes. Such systematic comparative treatment, which is attempted for the global commons regimes in Chapters 3–6, provides a necessary basis for consideration of regime effectiveness, incidence and change. The remainder of this chapter, thus, involves a consideration of how regime concepts can be utilized and more specifically how they can be adapted to describe and analyse the governance arrangements for the global commons.

Issue Areas

Regimes govern issue areas. This, according to Young (1989: 13), is what distinguishes them from the broad 'international orders' that superintend activities over a 'wide range of specific issues'. Regimes are:

> more specialized arrangements that pertain to well defined activities, resources or geographical areas and often involve only some subset of the members of international society.

Although we may separate out the principles, norms, procedures and rules that are specific to an issue area, they will necessarily relate to overarching international orders – free market capitalism or the legal and political order of state sovereignty. They will also, as was evident in the controversies over NIEO, reflect challenges to the existing order.

Having said this, it is the 'issue-specific' nature of regime type institutions that is their defining characteristic. The boundaries of a regime will be determined by perceptions of the extent and linkage between issues. Thus, the starting point of analysis must be the task of defining what constitutes an issue area. (There is always a danger of circular argument here particularly when there is uncertainty as to whether the conception of an issue area was antecedent to the growth of a regime although in the cases under consideration this was not usually so.) In the case of the global commons there is some basis in tangible reality and physical boundaries, but the key definitions are socially constructed. They may reflect powerful interests and values or even the established practices of organizations. A form of regime analysis that has been relatively neglected is the fundamental one of how agendas are set and issues arise, alter and are aggregated together. Who defines what this social construct – the issue area – will be? Keohane (1984: 61) provides an operational definition in which issue areas are 'sets of issues that are in fact dealt with in common negotiations and by the same and closely coordinated bureaucracies'. This is tantamount to saying that in the international system it is governments that decide issue areas, although it may well be that on closer inspection they reflect the interests and predilections of non-governmental groups and often, more evidently, the corporate sector. Also, as Brundtland (WCED, 1987: 310) observed, the structuring of issue areas is likely to be much influenced by the internal division of responsibilities within governments, which effectively serve, for example, to separate trade and environmental issues.

Issue areas will frequently not be coincident with the realities of physical and even socio-economic systems. They may actually be obstructive to the solution of commons problems. This mismatch between physical systems and issue areas proves to be a consistent theme across the various global commons. Outer space is treated as a set of discrete issue areas often overlapping terrestrial activities. There is a geostationary orbit (GSO) and frequency use issue area, but separate sets of issues concerning information flow, military uses and astronaut safety. The changing status of mobile commons resources has already been mentioned. Were fish to swim across the Antarctic Convergence (approximately 60°S) they would move from one issue area and regime to another (from the high seas to the area of the Antarctic Treaty). Whales, which most definitely do swim backwards and forwards across this line, constitute a completely separate issue area associated with the 1946 International Convention on the Regulation of Whaling.

In terms of regime effectiveness the construction of issue areas is a very significant matter. Natural scientists and the environmental movement increasingly adopt an holistic view of the commons, but the available governance mechanisms are partial and fragmented. There is a discontinuity between what may be termed the domain of the regime (a dictionary definition of domain is 'land governed by a ruler or government') and the physical and socio-economic dimensions of the commons in question. As will be argued in Chapter 7, the correspondence between the domain of the regime, determined by the construction of issues and best estimates of what would be required to ensure sustainable development or efficient and equitable management, has critical significance for a regime's effectiveness. A moment's consideration of the narrowness of attempts (in the Framework Convention on Climate Change 1992) to tackle the multifaceted problem of predicted global warming, would serve to make the point. Issue areas, as Keohane (1984: 61) points out, are unlikely to be static; they are 'defined and re-defined by changing patterns of human intervention: so are international regimes'. Thus, one of the greatest challenges to activists involved with environmental politics is to reconstruct the issues which must form the basis of regime development.

Actors

The description of issue areas presupposes a knowledge of the political actors and their interests. All too often it has been assumed that it is sufficient to restrict analysis to state governments. Such a view is often based on the assumption that governments actually do aggregate the varying demands and interests within society and are at the same time the only actors capable of exercising real authority over the behaviour that leads to the ruin of common resources. It also reflects the legal and organizational structures of the international system. However, it contradicts the findings of much of the International Relations literature of the last 20 years. This revealed the ways in which modernization and interdependence have weakened the extent to which state authorities can exercise control. Politics in the global system was no longer, if it ever had been, the exclusive preserve of nation states. Their role was now being supplemented and even challenged by a range of non-state and transnational actors.[3] One justification of the regime concept itself was that the traditional view of international cooperation, expressed in terms of formal organization and law, was no longer adequate to these changed realities (Ruggie, 1975). Critics were not slow to point out the irony whereby authors, previously noted for their desire to break free from a state-centric view of the world, relapsed into it once they came to consider the question of regimes.[4] To be fair, there has been a veritable explosion of interest in the non-governmental organizations (NGOs) and 'new social movements' phenomena during the 1990s along with a revival of attempts to conceptualize what Burton (1972) once termed world society and which is now described in terms of world or global civil society (Wapner, 1995). What is being referred to here is definitely not the international 'society of states' or 'international community' but webs of affiliation and political activity which cut across national boundaries and interpenetrate formal authorities at all levels but are not part of them. With the rise of green activism they have had a particular relevance for international environmental politics.

Nonetheless, state governments continue to be the formal participants in global commons regimes, often jealously guarding their status. In describing the politics of the regimes it is notable that there are long-standing coalitions of states which have some level of organization and usually concert their positions. The most prominent is the Group of 77 (G77). As a coalition broadly representing the developing countries of the South and constituting a permanent caucus at the General Assembly and elsewhere in the UN system, the G77 has played a significant role in the events described in this book. Now comprising over 120 members, it organized the campaign for a New International Economic Order, was a primary advocate of common heritage ideas, and continued to operate as a caucus throughout the 1992 Rio Earth Summit, and beyond.[5] Other coalitions, such as the Alliance of Small Island States (AOSIS) or the Geographically Disadvantaged and Landlocked States in the Law of the Sea negotiations, tend to arise in relation to specific issues and negotiations.

Such groupings may structure the interstate politics of the global commons but they do not constitute actors in the sense of entities with independent status and volition. The European Union (EU) most certainly can represent such an actor and is qualitatively different from other regional state coalitions. It is a very significant protagonist in its own right, particularly when issues within the ambit of the Treaty of Rome are negotiated. It has also acquired a distinct legal personality for itself and has its own special status as a Regional Economic Integration Organization (REIO), a category of participant of which the EU is the only extant example, but which is now routinely mentioned in international environmental agreements. The respective competencies of member states as opposed to the Commission has been a difficult and often confusing matter in global commons negotiations.[6]

The United Nations has usually provided the arena for the negotiation of global commons issues, whether in the General Assembly itself, in special conferences or in the Specialized Agencies. But there is also a sense in which the secretariats of some organizations and programmes, notably the United Nations Environment Programme (UNEP), may be seen as significant actors in their own right. Set up by the 1972 Stockholm Conference, UNEP has defined its mission in terms of a 'catalytic and coordinating' role across the UN system and has been heavily involved in the stratospheric ozone layer negotiations, the development of marine pollution management and the preparations for the 1992 UNCED (although it was specifically excluded from the negotiation of a framework climate treaty). Working within the UN system, a number of key individuals may also be identified as actors; defining issue areas and, indeed, providing a personal link between them. Maurice Strong, who chaired the 1972 and 1992 UN Environment Conferences and was the first Executive Director of UNEP, Mostafa Kamal Tolba, Strong's influential successor at UNEP, and Tommy Koh of Singapore who was prominent in both the Third UN Conference on the Law of the Sea (UNCLOS III) and UNCED, immediately spring to mind in this regard.

Private sector business organizations and above all large transnational corporations are actors of the first importance. It is, after all, their activities, whether in using CFCs, in launching satellites into geostationary orbit, or in energy use, which frequently constitute the ultimate object of international cooperation to preserve the global commons. The extent of corporate involvement is often underestimated as analysts focus upon the campaigning activities of NGOs, but their lobby groups, such as Business Council for Sustainable Development in the UNCED process, are well financed and represented. It is often difficult to separate the interests and activities of states and corporations. (There are some important actors, such as Intelsat and Inmarsat, which as international common user organizations having both private and public sector members are impossible to classify.) The Montreal Protocol negotiations on the stratospheric ozone layer exhibited the intermeshing of state and corporation in full measure. Government policies on occasion appeared to represent little more than a reflection of the commercial perspectives of Du Pont, Atochem or ICI. In a revealing but hardly impartial aside, Benedick (1991: 78) notes that an EC position paper during the negotiations was inadvertently distributed by the French delegation on Atochem headed note-paper. Often what appears, at first glance, to be a 'national' delegation appears, under closer scrutiny, to be composed mainly of representatives of the private sector. A good example is provided by the composition of delegations to ITU World Administrative Radio Conferences (WARCs). At the 1979 General WARC, that set up the orbit conferences of the mid-1980s, the US delegation's 65 members were mainly drawn from a whole range of private sector concerns. This was probably an extreme example born of the uniquely pluralistic system of the United States. Private sector participation in national delegations exists elsewhere but is often less overt and more difficult to establish. Such identification is complicated by the fact that in the organization that Jacobson (1974) christened a 'Potpourri of bureaucrats and industrialists', personnel often appear to move smoothly between private boardroom, national delegation and international secretariat. This does not need to imply some from of corporate conspiracy theory. What it does demonstrate is that in many of the areas surveyed, the issues are so technical and specialized that governments rely on external assistance which may come from business or NGOs. This provides an interesting external parallel to the well established analysis of interest group access and influence in domestic government.

The development that has received most attention in global environmental negotiations has been the rise in the number and importance of NGOs. NGO activities provide the empirical basis for conceptions of an emergent global civil society. They now routinely enjoy observer status within UN organizations and conferences (Morphet, 1996) and may, on occasion, be invited to send representatives to form part of national delegations. (This has been the case with UK delegations to recent International Whaling Commission Meetings and also with the UK delegation to UNCED.) NGOs vary greatly in scale, approach and function. Greenpeace, Friends of the Earth (FoE) and the World Wildlife Fund (WWF) are rather different organizations with distinct strategies for exercising influence. Some of them deploy a level of scientific and technical expertise unmatched by all but the largest developed states, others are capable of mobilizing consumer boycotts or are prepared to engage in direct action or local community building. To give some idea of the range of NGO operations, at the 1981 meeting of the International Whaling Commission 52 NGOs were represented ranging from the American Cetacean Society to the World Council of Indigenous Peoples, by way of the Assembly of Rabbis, Greenpeace and the Royal Society for the Prevention of Cruelty to Animals (Birnie, 1985: 752–757).

As well as providing a transmission belt between the concerns of domestic publics and international negotiations, 'linking the global and the local', the NGOs have become increasingly significant in making connections across the various issue areas. To take the example of Greenpeace International; it is an organization with a world-wide membership of 6.75 million and an annual revenue, in 1990, of in excess of $100 million (Princen & Finger, 1994: 2). With a well defined and militant stance on a range of commons issues it has played a prominent role in the whaling and Antarctic regimes and has been active on questions of maritime pollution and climate change. There have also been extensive interconnections between the swelling number of NGOs. This too was, in part, a consequence of Stockholm and the creation by UNEP of the Environmental Liaison Centre in Nairobi. Now named the Environmental Liaison Centre International (ELCI) it had 726 members in 1993 (Princen & Finger, 1994: 2). It also maintains connections with over 8000 NGOs worldwide (Thomas, 1992: 28).

Organizational links can only describe part of the story. There is also the important role of change in that common scientific understanding which underlies the principles of many regimes. The generation of scientific consensus may be formally organized as with UNEP's work on ozone trends, the World Climate Conferences of the late 1980s and most notably the Intergovernmental Panel on Climate Change (IPCC). Often underlying such efforts are more amorphous 'epistemic communities' difficult to define as actors but nonetheless significant. These 'transnational networks of knowledge based communities' (Haas, E., 1990: 349), usually comprising technical and scientific experts, have received a great deal of recent attention in the literature on international cooperation. Some analysts regard them as central to an understanding of the 'cognitive' bases of regime formation. For our purposes, possible epistemic communities exist amongst Antarctic scientists, international maritime lawyers, radio spectrum and telecommunications engineers (who tend to be the dominant group in ITU), as well as the more well known expert groups associated with the ozone layer and climate change issue areas.

There has been relatively little study of the processes whereby the various actors have engaged in agenda setting or the manufacture of issue areas. Most mysterious and intangible, but in a way most significant, are the marked shifts in public opinion involving popular disquiet over issues such as ozone depletion and climate change and different valuations of the environment. The NGOs have clearly had a significant hand in this process, and one of the most developed analyses of their role (Princen & Finger, 1994) argues that their significance lies not just in linking the local and the global but with connecting the 'biophysical to the political'. In this they were assisted by a number of high-profile 'shocks' such as the discovery of the Antarctic ozone 'hole', the Chernobyl accident and the *Exxon Valdez* oil spill.[7] At one point in the late 1980s developed world leaders were actively vying with each other to establish their credentials in respect to such public concerns. Popular anxieties and causes helped to define new issue areas and percolated through to the principles and norms of existing commons regimes.

Principles and Norms

These, according to standard academic usage in International Relations, are the key characteristics of a regime. When they change the regime changes, but rules and decision-making procedures do not enjoy such prominence. However, it is quite possible to dispute the distinction between principles, norms and rules and to claim that there are only rules: some constitutive of the institution and others having a more specific operational content. Ostrom's (1990) work on the governance of domestic commons does this very convincingly. Donnelly (1986) has also argued that the distinction between principles, norms and rules is of little analytical value and merely refers to rules of greater or lesser specificity. He adopts a simple two-fold categorization of norms and decision-making procedures.[8] These observations have validity, but the temptation to re-work the standard categories has been avoided here simply because the Krasner (1983) regime definition is still serviceable, as long as the meanings and interrelationships of its elements are properly specified and, above all, because it enjoys such wide currency that it makes sense to conform.[9] The importance of being able to utilize standard terms that facilitate comparison with other studies outweighs any conceptual misgivings.

'Principles' and 'norms' are often used synonymously. One definition of a principle, widely used in ordinary speech, refers to standards or rules of personal conduct. At the same time norms can denote valued, socially established or even merely average behaviour. Rigid adherence to Krasner's categories provides a somewhat arbitrary escape from this terminological morass.

Thus, principles are defined as 'beliefs of fact causation and rectitude'. Norms are 'standards of behaviour, defined in terms of rights and obligations'.

One of the virtues of a comparative analysis of the development and characteristics of related regimes is that their interconnections may be clarified as well as the influence of changes within the broader international system. Although regimes are 'issue area specific' their norms and principles do not exist in a vacuum. As the preceding treatment of the various actors in global commons negotiations demonstrated, many of them have consciously attempted to make connections between their activities in various issue areas. The North–South dialogue and the campaign for a NIEO, articulated by the Non-Aligned and G77 nations, evidently spilled over into the discussion of principles and norms in a number of the regimes under consideration.

The 'law-making' activities of general UN Conferences and UN General Assembly resolutions have also been significant as providers of norms and principles for a number of regimes. Maritime regimes, for example, and even in part the Antarctic regime, may be regarded as being derivative of the general Law of the Sea as enunciated in the Third UN Law of the Sea Conference 1973–82. Also, the landmark UN environment conferences at Stockholm and Rio provided authoritative statements that were intended to influence and also to reflect the specific norms and principles of particular regimes. A way of describing these connections is to assert that regime norms and principles, although specific, are far from unique and are 'nested' within a hierarchy.[10]

Principles

'Beliefs of fact and causation' would include those shared scientific under-standings, as to the nature of the physical world, upon which many commons regimes rest. Examples include the chemistry of stratospheric ozone depletion by the action of CFCs, fish population dynamics, and the properties of the GSO and the frequency spectrum.

The period between Stockholm and Rio witnessed changes of great magni-tude in the scientific understanding of environmental change. In part these were directly stimulated by the Action Programme of the 1972 UNCHE which, amongst other things, set up UNEP. Scientific investigations, organized by UNEP, into the depletion of the stratospheric ozone layer and the global warming and climate change hypotheses (see Chapter 6 below), led to a major shift in comprehension which was to inform the principles of a number of regimes and to stimulate the creation of others. While Stockholm had mainly been concerned with problems of transboundary pollution and measures to 'clean up the mess' left by industrialization, the problems confronting (but hardly solved by) UNCED were generally perceived to be of an altogether different scale and character. The new underlying belief, reflected in the public debates and World Climate Conferences of the late 1980s, was that global environmental change was occurring and that it was now the responsibility of the world community to deal with the causes. The effect was to create a completely new issue area and then an embryo governance regime for green-house gas emissions.

As outlined in Chapter 1, the most significant principles associated with commons regimes are those that actually define their status in terms of property rights and, in the case of common sinks, responsibilities. Within these basic categories there are also important and, often disputed, allocative principles. Open access implies a 'first come first served' principle but this can also apply to more regulated common property resource (CPR) regimes where the rights of initial exploiters are actually registered and protected (in the ITU *a priori* rights vesting). Market principles can be applied to CPR regimes as well as centralized forms of allocation and resource planning. There are parallel arguments about the relative merits of 'command and control' as opposed to markets in 'rights to pollute' in relation to the development of regimes for common sinks. The 'common heritage' concept is often seen as the apotheosis of a central planning in relation to the commons – an impossibly 'socialistic' enterprise in which 'political' demands for equity take precedence over 'efficient' market allocation of resources. This was to be the Reagan administration's criticism of the principles of the embryo seabed regime, yet, as will be discussed in Chapter 3, it still contained significant market-based elements. A related, but not entirely similar, competition between principles has been the central feature of the recent development of the ITU orbit/spectrum regime and debates about the planning of a common pool resource, either by *a priori* rights vesting which legitimates a situation of 'first come first served' or a more planned system of *a posteriori* rights vesting. Discussion of climate change control regimes exhibited arguments over the relative merits of essentially national pursuit of emission reduction targets or the planned creation of a market in rights to pollute – a tradeable permits system (Grubb, 1992). The Kyoto Protocol 1997, while committing developed nations to emission reduction targets, also provides a range of potential mechanisms, including emissions trading, for their achievement. Clearly, rival allocative principles involve beliefs about causation and 'facts', but they are also most definitely pervaded by beliefs about 'rectitude' or, to put it in a less stilted manner, the conflict between political and economic value systems.

The arguments about allocative principles within specific regimes, from the late 1960s through to the mid-1980s, cannot be seen in isolation from the general campaign for a NIEO, which represented an explicit challenge to the assumptions upon which many such principles were based. Krasner (1985: 115) has spoken of the 'coherence of the NIEO programme across a wide range of issue areas'. Most significantly Arvid Pardo's introduction of the concept of the oceans as the 'common heritage of mankind' in 1967, seven years before the

UN declaration on the economic rights and duties of states, constituted an important part of the emerging campaign for a New International Economic Order. Ocean resources should not be open to free exploitation under a market system, but equitably managed for the collective benefit. During the 1970s, NIEO-related campaigns arose in UNESCO and ITU. These, which were explicitly connected to the North–South debates in the General Assembly, were summarized in terms of the New World Information and Communications Order (NWICO) and, as we shall see, were a specialized application of NIEO principles to specific issue areas relating to news and broadcasting flows and equitable access to, and distribution of, the orbit spectrum resource.

The link between environmental concerns and the developmental objectives of the South achieved global prominence with the convention of the aptly named United Nations Conference on Environment and Development (UNCED) in 1992, although the connection had been a contentious item at Stockholm in 1972. Notwithstanding, the politics of development and the environment ran along parallel tracks for the rest of the decade. It was only after the virtual collapse of the 'North–South dialogue' that they were brought together again by the Brundtland Commission. Set up by the UN General Assembly and reporting in 1987, the Commission coined the concept of 'sustainable development'. Briefly stated this meant 'development that meets the need of the present without compromising the ability of future generations to meet their own needs' (WCED, 1987: 43).

Clearly, there may now be widespread (but by no means complete) agreement that unrestrained development and utilization of resources (particularly of the global commons) will lead to serious resource and survival problems if not the collapse of the biosphere. The norm that all economic development ought in future to meet the test of sustainability follows from this. The arguments at UNCED demonstrated how contentious the concept of sustainable development can be, raising as it does central issues of relative wealth and patterns of consumption and exploitation. However, it is difficult to imagine future environmental or commons agreements that do not incorporate sustainable development into their principles and norms.[11]

In some respects the 1992 Rio Declaration on Environment and Development is a little more specific. It acknowledges both the 'different contributions to global environmental degradation' and also that 'States have common, but differentiated responsibilities' (Principle 7). Many commentators have traced a critical connection between the insatiable demands of the world trade and monetary order and the unsustainable destruction of natural resources by LDCs constrained to conform to the disciplines of the IMF. The Rio declaration, however, contains a statement of faith in the open world market system: 'States should cooperate to promote a supportive and open international economic system that would lead to economic growth and sustainable development in all countries' (Principle 12). As a statement of 'fact, causation and rectitude' this may appear to be a somewhat implausible 'squaring of the circle' of economic growth and sustainability, but its acceptance (and inclusion in the FCCC and doubtless future agreements) demonstrates the extent of change since the days of the NIEO.

Principles of rectitude involving equitable treatment and the relationship of the developed to the less developed world have recently been joined by those which value the environment, not as a potentially exploitable resource to be fairly distributed, but for itself. Thus in the evolution of the whaling regime and the Antarctic Treaty System it is possible to perceive the incorporation of principles embodying so-called 'wilderness values'.

'Wilderness values' can be a radical extension of traditional interest in the conservation of habitats and wildlife. Certain forms of green political thought now adopt an avowedly 'ecocentric' rather than 'anthropocentric' view of the world. This has led to controversies over the relative valuation of the preservation of a 'wilderness' and human development.[12] It is easy to dismiss this as the insignificant theorizing of a political fringe, but there is evidence in some of the issue areas, notably Whaling and Antarctica, that it is beginning to mean more than this.

Norms

As must, by now, be fully evident, the distinction between principles and norms is often difficult to draw. Norms defined in terms of standards of behaviour, rights and obligations, often stem directly from principles. Hence an obligation placed upon states not to allow their nationals to engage in the commercial hunting of whales is the operative extension of principles concerning the necessity to conserve this form of marine life. It is difficult to describe 'sustainable development' as a principle or a norm, for it is in effect both. Matters are further complicated when such a widely used concept as the 'precautionary principle' turns out to be a norm.

Norms often involve rights to use the various commons, for example the right to fish or navigate the high seas or to use the radio frequency spectrum free of harmful interference, coupled with injunctions not to infringe the enjoyment of such rights by others. They can involve prohibitions of certain generally harmful activities – the emission of CFCs, certain forms of maritime pollution or the militarization of common spaces such as Antarctica or the deep seabed. Once again, however, these norms rest upon an established principle that such activities are harmful and contrary to the common interest. The Stockholm Conference produced a particularly important statement of norms relating to environment and development which proved to be influential in a number of specific regimes. Inevitably it was labelled a principle (Principle 21). Adroitly combining responsibility for transboundary pollution with assertion of the economic rights of LDCs to exploit their own national resources, it set out:

> the sovereign right [of states] to exploit their own resources pursuant to their own environmental policies and the responsibility to ensure that activities within their jurisdiction or control do not cause damage to the environment of other States or areas beyond the limits of national jurisdiction (Principle 21, Stockholm Declaration, 16 June 1972)

At a rather more mundane, but nonetheless important, level UNCED registered the emergence and spread of norms relating to the practice of environmental protection which were unheard of, or at least little known, in 1972. In essence these were developed by environmental lobbies, scientists and policy-makers in the advanced industrial countries. They came to be inserted into existing global commons regimes during the period and are neatly summarized and legitimated in the Rio Declaration (many practical applications are worked out in great deal in Agenda 21 which was by far the weightiest document, in every sense, produced by the Rio Conference). They include principles 15–17 of the Rio Declaration (which would in our terminology constitute norms because they are injunctions). First there is the precautionary approach, which stipulates that 'Where there are threats of serious or irreversible damage, lack of full scientific certainty shall not be used as a reason for postponing cost-effective measures to prevent environmental degradation' (Principle 15). Then there is the stipulation that the polluter should 'bear the cost of pollution, with due regard to the public interest and without distorting international trade and investment' (Principle 16). Finally, there should be environmental impact assessment for intended activities 'likely to have a significant adverse impact on the environment' (Principle 17).

Decision-making Procedures

Both domestic small-scale regimes as well as those for the global commons require some mechanism for the making of collective choices. Such choices may involve day-to-day application of the rules (the quantity of fish that may be caught or who has the right to draw water from an irrigation system at a particular time) or much more extensive consideration of changes in the rules themselves and even the norms and principles that define the regime.[13]

In reviewing the institutional arrangements for the global commons, it soon becomes evident that there are a range of collective choice mechanisms varying in their formality and organizational complexity. The most *ad hoc* procedures involve periodic meetings of the parties (for example Antarctic Treaty Consultative Parties Meetings) who may not even have the services of a dedicated secretariat. At the other end of the spectrum there may be a set of elaborate international organizations as, for example, those designed for the administration of the seabed regime at UNCLOS III or the complex architecture of the International Telecommunication Union that superintends the orbit/spectrum regime. (Again there is some terminological confusion, because it is commonplace to refer to international organizations as institutions; hence the IMF and World Bank are known as 'the global financial institutions'. The word institution has a particular meaning in this book, corresponding to standard sociological usage, and thus it will not be employed to describe formal organizations.)

Systematic, although not always successful, efforts have been made to coordinate environmental policy across the system and the division of responsibilities between a whole range of UN organs has long been regarded as a major problem. It has been addressed by the System Wide Medium Term Environmental Plan (SWMTEP) and the creation of Designated Officers for Environmental Matters (DOEMs) but above all through the coordination activities of UNEP. The organizational capacities of UNEP are in fact shared by a number of regimes. It provides the secretariat for the ozone regime and has special functions in the area of maritime pollution – although it was not chosen to provide a secretariat for the climate change regime. The dense thicket of inter-organizational relations relevant to the global commons is well beyond the capacities of this book to describe – the way in which, for instance, WMO, UNEP, UNESCO/IOC, FAO and ICSU were all involved in the sponsorship of studies of the atmospheric commons and climate change (Second World Climate Conference 1991). The essential point is that, as with norms and principles, the organizational components of particular regimes cannot be seen in isolation. A significant instance is provided by regimes that 'share' an organization. The International Maritime Organization (IMO) acts as a secretariat for a number of regimes, just as ITU provides the organizational basis for space and terrestrial radio communications along with telecommunications by cable.

It is difficult to be dogmatic about the types and levels of organization that might be effective. Much will depend upon whether 'self monitoring and enforcement' is regarded as more acceptable than the creation of some central body to perform such functions. A dedicated organization may not be required, or when a new regime is created an existing organization may be persuaded to expand its functions. Such matters can be politically contentious and engage various organizational interests within the UN system. Salient examples are provided by the refusal of the parties negotiating the FCCC to allow UNEP to provide the secretariat and by discussions within the Montreal Protocol regime on the desirability of allowing the Global Environmental Facility of the World Bank (GEF) to administer the compensatory Ozone Fund.

A plausible generalization might be that the greater the number, frequency and complexity of decisions that have to be made, the greater is the requirement for permanent organizations. As the scope of regimes expand there seems to be a tendency in this direction; witness the increasing level of organization even in the Antarctic Treaty System, which had prided itself on its rather *ad hoc* procedures. This trend has been resisted, particularly by Northern governments that see themselves as shouldering the financial burden of new organizations. Thus, the US and UK went into discussions preparatory to UNCED and the FCCC with the declared policy of 'no new organizations'.[14] One result of the Rio Conference was, however, to set up a new UN organ, the Commission for Sustainable Development, specifically charged with coordinating the activities of existing organizations and programmes.[15]

An issue that will inevitably recur in discussions of international-level decision-making, both academically and in terms of practical politics, is that of voting rights and procedures. The formal parties are sovereign states and the usual practice within the UN system has been a system of 'one state one vote', where majority decisions are, however, not always regarded as binding (as in General Assembly Resolutions). Majority voting procedures were the procedural basis upon which the Group of 77 was able to assert its demands for a NIEO. As has been repeatedly pointed out by the developed countries, such majoritarian procedures fail to reflect the stake that particular states may have in an issue under discussion or their financial contribution. The IMF and World Bank provide an example of the alternative system whereby votes are allocated according to financial contributions or quotas. If voting rights do not align with the power of the various participants, it may be argued that organizations are rendered impotent. Majorities may be constructed for resolutions, but if they do not include, but rather offend, powerful players they will be of little consequence. Many of the regimes under consideration have grappled with these issues and have devised quite complex voting arrangements which vary according to the rules or issues under discussion and seek to protect the interests of various groups of interested states and associated private sector corporations. A good example is provided by the complex provisions for representation and voting on the Council of the International Seabed Authority. These matters are mentioned here in outline because they have some prominence in the case studies. They will also be considered in greater depth in the context of regime effectiveness in Chapter 7, and in Chapter 8 which deals with the politics of regime creation and change.

Rules

According to Krasner's (1983) definition these are to be regarded as 'specific prescriptions or proscriptions for action'. They constitute particular applications of general regime norms and principles. Thus, for example, rules on distribution and compensation would follow from the common heritage principle, if that were to be accepted. In practice, however, as Kratochwil (1989: 59–60) has pointed out, the relationship between the two is rarely as hierarchical or simple as Krasner's distinction may imply. This is a continuing source of difficulty which has significant implications for the identification of regime change.

Rules are usually codified in formal multilateral legal agreements (indeed as we have seen Porter & Brown (1991) regard such instruments as the defining characteristics of a regime). Yet they can have a softer and more informal character and any institution is likely to develop a set of understandings and accepted practices that supplement its more formal rules. Accepting this to be the case is one thing, but, as much of the regime literature (including the case studies in subsequent chapters of this book) demonstrates, actually identifying such informal rules is quite another.

Beyond referring to 'prescriptions and proscriptions' Krasner and his associates provide little guidance as to how rules are to be categorized. Categorization will necessarily reflect the interests of the analyst, but the most obvious way to proceed is to consider how different types of rule relate to the functions of the regime. A focus on the governance of the commons can provide a clear sense of direction here. As mentioned in the introductory chapter there is a substantial literature on the institutional requirements for the effective governance of small-scale commons. There should, for example, be a clear specification of rights and responsibilities to a common resource coupled with effective monitoring and compliance mechanisms (Berkes, 1989; Ostrom, 1990). Despite the manifest differences in scale there are enough essential structural similarities to make domestic experience at least a relevant starting point for the analysis of the type and functions of rules at the international level. It is argued that, as well as possessing some mechanism for ensuring collective choice, a commons governance regime is likely to have all or most of the following rule-related functions: standard setting, distribution, information, enforcement and knowledge generation.[16] This provides a basis for the following classification of regime rules:

1. *Standard setting*, involves promoting desirable actions and prohibiting others. It therefore covers the whole range of rules for environmentally beneficial behaviour in the Antarctic or in the utilization of the oceans or atmosphere. It also denotes 'technical standards' that are required, for example in the utilization of radio frequencies such that the resource may be efficiently enjoyed by all.
2. *Distribution*. This function requires rules and procedures for the allocation of shares or user rights to a common resource, plus obligations in terms of provision and renewal. Basic to any commons regime, they are graphically illustrated by the catch 'quotas' allowed by the whaling regime or more recently by arguments about national shares of total

carbon dioxide emissions. A further function of considerable importance, which may well determine the participation of less well endowed actors, relates to compensatory payments or technology transfer.

This was the most prominent issue in discussions of the seabed and has a central place in pre- and post-Rio discussions of the environment–development nexus. If a resource is to be defined as a 'common heritage', as the deep seabed was, the question of what revenues and benefits would be payable to those who were not in a position to mine its resources becomes critical.

3. *Information.* Collective action is dogged by the 'free rider' problem. Commitments to abide by collective decisions will be conditional upon adequate assurances that others are doing likewise. Thus reporting and monitoring arrangements and transparent information sharing are likely to be the *sine qua non* of a workable regime. All the regimes under consideration have struggled with this problem and for some, such as the stratospheric ozone regime, it is a key determinant of effectiveness.

4. *Enforcement/compliance*, is closely allied with the monitoring function. Excessive attention has been paid to the imposition of formal sanctions against members or outsiders, yet the whole 'governance' problem in international relations and also as specified in discussions of domestic commons self-management, stems from the undesirability or impossibility of central coercive authority. Mechanisms are, therefore, likely to have a more subtle self-regulatory character. Amongst sovereign states there will be reliance upon threats of reciprocity and horizontal enforcement, which may rest heavily upon a national concern to preserve international status and reputation. The relative absence of central monitoring and enforcement places a heavy reliance upon the governments party to the various commons regimes to enforce compliance by their own subjects, whether they be misusing orbit and frequency spectrum, illegally hunting whales or continuing to employ prohibited substances under the Montreal Protocol.

5. *Knowledge.* At the highest level scientific discoveries may, as in the case of the stratospheric ozone and climate change, define new issue areas and serve to trigger the initiation of regimes. At a more mundane but very significant operational level, the development of scientific and other knowledge will be a major determinant of the rules that are made and the possibility of their implementation. Thus, rules for common fishery resources (for example in the Antarctic regime) depend upon ongoing investigations into the nature of marine ecosystems and population dynamics. The identical point can be made about rule-making activities across virtually all the cases under consideration. Despite the widespread adoption of 'precautionary' norms it is equally the case that a lack of scientific evidence remains the most popular rationale for avoiding burdensome regulation to conserve stocks or protect

the environment. In terms of climate change there has been a lively and very political debate about the validity of predictive general circulation models. The elusive 'holy grail' of policy remains the 'win, win' or 'no regrets' solution which will render a politically and economically cost-free answer to the problem of sustainable growth.

For all these reasons, most of the regimes under consideration have some capacity and associated rules and procedures for the generation and/or evaluation of relevant knowledge. Some, such as the ozone regime, have developed a complex set of knowledge processes designed to determine the development and modification of rules, whereas at the other extreme the International Whaling Commission existed for years with hardly any capacity for understanding the effects of whaling or properly estimating what a sustainable catch might be.

Regime Change

Much of the academic discussion of regimes has centred upon the question of the explanation of change. This may involve the creation of a new regime where none had previously existed, but it has most prominently concerned the alteration of norms and principles. It should be recalled that the development of theory about regimes occurred within a specifically American policy context. The precipitate changes occurring during 1971 in the world monetary regime, involving the collapse of a system of fixed parities anchored to the dollar, provided the problematic of how order was to be restored in the apparent absence of US dominance. The attempt to understand changes and the circumstances in which regimes might be created inevitably involved reliance upon the established theoretical traditions in International Relations and International Political Economy.

It was, therefore, probably no accident that the dominant thesis on regime change, 'hegemonic stability', turned out to be an intellectual descendant of realism. This thesis attributed the rise and maintenance of regimes to the power of an hegemonic state. Examples of the imposition of rules by the powerful were provided by the 'pax Britannica' of the nineteenth century, sustaining the freedom of the seas, and in the twentieth by the American role in creating the post-war economic settlement that underpinned the sustained growth of the world economy over the next quarter century. This view has not met with uncritical acceptance. Neither, despite a number of attempts, has it received conclusive empirical verification (Webb & Krasner, 1989). It did, however, capture the sense of (probably exaggerated) American alarm in the wake of Vietnam and the traumatic monetary events of 1971. It also shares some key assumptions and literature (in relation to the public goods problem) with the 'Hobbesian policy idiom' applied to domestic commons.

Alternatives, within a broadly liberal and utilitarian tradition have been provided by a number of scholars. Most well known is Keohane (1984) who, under the apt title of *After Hegemony*, suggests that microeconomic theories, of the type which have already been seen in relation to the definition and regulation of common property resources, can be applied to the creation and sustenance of international regimes. As with the state in the domestic arena, the function of regime institutions in the international system is to cope with 'market failure' and minimize transaction costs.

More firmly rooted in what has been called the 'Grotian' tradition, emphasizing the pervasive importance of international institutions in determining behaviour, is the extensive work of Young (1977, 1982, 1983, 1989, 1994). There is also a significant body of writing on the nature and determinants of cooperation, often utilizing game theory, which has direct and often explicit relevance to the study of regime creation and change.[17]

A third general approach to explanation is also identifiable, which is founded not upon power or interest assumptions about behaviour but on the significance of cognition and consensual knowledge. This 'cognitive' writing is associated with the functionalist approach to international organization. Transnational collaboration between scientific and technical experts (who may relate to each other as an epistemic community) has had such evident significance for the development of environmental and other knowledge-dependent regimes that cognitive approaches appear eminently suited to their analysis. In this regard Peter Haas' (1990b) study of the development of the Mediterranean anti-pollution regime has been particularly influential.

A consideration of how these approaches fare in explaining the creation and development of the various global commons regimes forms the substance of Chapter 8. As they have concentrated on the question of regime change, it is important to establish exactly what this denotes. Change is not as straightforward an object of study as it may at first appear. The standard delineation of regimes, used throughout this chapter, is unambiguous on the matter:

> In sum, change within a regime involves alteration of rules and decision-making procedures, but not of norms or principles; change of a regime involves alteration of norms and principles; weakening of a regime involves incoherence among the components of a regime or inconsistency between the regime and related behavior. (Krasner, 1983: 5)

This definition of change follows directly from the distinction between norms and principles on the one hand and rules and decision-making procedures on the other. Because the former are seen as determining the latter, it must be the case that *significant* change is change at the superior level of norms and principles. This simple distinction is not, alas, unproblematic when applied to cases. Regimes are likely to be in a constant state of alteration. There are probably rather few in which it is possible to establish an abrupt shift of norms and principles similar to that which occurred to the Bretton Woods regime in the early 1970s. The Montreal Protocol regime for stratospheric ozone is, given the sheer volume of coverage, fast approaching the kind of status enjoyed by the monetary regime in academic discussions. It, too, may be rather atypical in that it was specifically negotiated from 'first principles', which have enjoyed a consistent hierarchical relationship with the specific, and frequently amended, rules on particular ozone-depleting substances.

An excessive concentration on these two cases may obscure what may be more commonplace processes of messy incrementalism.[18] Important changes may hinge on a series of rule alterations and procedural shifts – the full impact of which can only be seen in retrospect. Anticipating the material in Chapter 4 below, which outlines a very significant change in the purposes of the Antarctic regime from 1959 to the present, one may legitimately ask: when did its principles and norms shift decisively towards conservationism? The answer appears to be that it was a slow process that can only really be tracked in terms of an accretion of sometimes quite minor rule changes and expansion of scope. The latter is illustrative of another important omission in many discussions of regime change. A concern with the nature of norms and principles may overlook what may be considered equally significant changes in the domain to which they apply.

In the end the question of regime change may be resolved quite simply. Instead of trying to pursue some universally acceptable definition of change it is far more sensible to conclude (with Kratochwil) that regimes may have several purposes. Criteria as to what constitutes significant change will, accordingly, vary in relation to the subjective interests of the observer. Thus, 'Whether we perceive a change of regime or within a regime much depends upon our perspective and purposes' (Kratochwil, 1989: 60). Bearing this in mind, discussion of regime effectiveness looms even larger in significance. The relevant questions about regime change become: how can a non-regime situation or a weak regime be altered into an effective form of commons governance? Under what circumstances do such institutions develop and decay, and how can we account for differences across various issue areas?

The argument in the remainder of this book attempts to grapple with these questions. The next four chapters attempt a regime analysis, identifying issue areas, regimes and their components in the four global commons, the oceans, Antarctica, space and the atmosphere. They are treated separately, but they share important common features and connections which are natural, legal and organizational as well as being located within what has been defined as a wider international order. Chapter 7 explores concepts of effectiveness in terms of a comparative study of the characteristics of the regimes and what is known of their performance. It should provide some basis for establishing those significant changes and variations between regimes that require explanation. It is only at this point, in Chapter 8, that what have been termed 'theories about regimes', whether power, interest or cognition based, are deployed in the discussion of the politics of regime creation and change.

Notes

1. Strange (1983: 354) has argued, in a trenchant critique, that a proper understanding of 'who gets what' in the international political economy 'is more likely to be captured by looking, not at the regime that emerges on the surface, but underneath at the bargains on which it is based'.
2. Kratochwil (1989) goes much further than this in arguing that norms in international relations actually provide the reasons for actions and are central to the decision-making process.
3. Works in this vein are: Keohane & Nye (1973) *Transnational Relations and World Politics*, Morse (1976) *Modernization and the Transformation of International Relations*, and Mansbach *et al.* (1976) *The Web of World Politics*. A well known rebuttal is provided by Miller (1981) in *The World of States*.
4. This observation is part of Strange's (1983) critique. The return of state-centric analysis may partly be explained by the need to make simplifying assumptions, something that is particularly marked when game theoretical approaches or those derived from microeconomic theory are employed.
5. The G77 came into being in the context of the 1964 United Nations Conference on Trade and Development (UNCTAD I), which was itself organized into various negotiating groups. G77 is not an organization but a caucus, and in many ways its regional groupings (Latin American, African and Asian) have a stronger identity. Membership overlaps to a significant degree with the Non-Aligned Movement which has worked in tandem with the G77 especially in the campaign for a NIEO.
6. The respective position of the Member States and the Commission is particularly complex when the EU engages in international environmental negotiations, because mixed competence pertains. Thus, depending on the issue, either the Commission or the Presidency will lead and there may be 'coordination' problems which complicate negotiation. Nonetheless, despite its internal complexities the EU has managed to play an increasingly prominent leadership role. On this and the concept of actorness in general see Bretherton & Vogler (1999). On the specifics of the EU as environmental negotiator see Vogler (1999).
7. Thomas (1992) provides an analysis of these events. Young (1989, Ch. 9), emphasizes the importance of seizing the political opportunities that such events and consequent public interest provide.
8. Ostrom (1990: 51–57) simply talks about institutional rules at the operational, collective choice and constitutional levels in order of ascending priority. 'Norms of behaviour reflect valuations that individuals place on actions or strategies in and of themselves not as they are connected to immediate consequences' (Ostrom, 1990: 35). Norms are seen not as a defining characteristic of an institution but as an important external variable that will determine the degree of self-interested opportunism and the degree of monitoring and enforcement that will be required. Keohane (1984: 57) discusses but rejects the concept of norms involving moral claims that transcend self-interest because of the utilitarian basis of the thesis he advances in *After Hegemony*. Donnelly (1986: 603 fn.12) would concur with Ostrom that in effect Krasner's categories refer to rules of greater or lesser specificity.

However, in the 'interests of clarity' he defers in part to Krasner's authority and uses the 'relatively neutral term "norms" to refer to the full range of a regime's normative principles (in contrast to decision-making procedures)'.

9. A well known addition to the Krasner definition was the suggestion by Rittberger (1993: 10–11) that 'effectiveness' and 'durability' should also be regarded as defining characteristics.

10. This is analogous to Ostrom's (1990: 51–53) conception of a nested hierarchy of rules in domestic small-scale commons regimes.

11. Sustainability does not yet have the full status of a international legal norm but, in the words of one legal authority, 'it is a notion around which legally significant expectations regarding environmental conduct have begun to crystallise' (Handl, cited in Birnie & Boyle (1992: 5)).

12. For a good review of the relevant literature from an ecocentric rather than anthropocentric perspective see Eckersley (1992) and especially pp. 39–42 on wilderness values.

13. The reader may be curious as to why 'decision-making procedures' have been placed before 'rules' in this discussion. This is simply because it was found that in analysing the cases it was desirable to have an initial understanding of key procedures and decision-making bodies in order to make sense of what was contained in the rules.

14. There is also the interesting case of recent ITU reforms involving the IFRB where the availability of new technologies has made it possible to put the Board on a part-time basis.

15. On this and the functions and development of UNEP see Imber (1993).

16. Somewhat different formulations of the kind of rules that will be found in successful commons regimes are to be found in Ostrom (1990: 90) and Berkes (1989: 26). Both emphasize the allocation of rights, collective choice arrangements, monitoring and enforcement.

17. For a substantial review, see Milner (1992).

18. Kratochwil (1989: 59–60) explores this point. He argues that regimes are usually the result of accretion and incremental choices and that even explicitly negotiated regimes often serve several purposes.

3

The Oceans

The oceans were the original global commons, fished and navigated for millennia. A systematic attempt to draw up principles for the use of the oceans is evident even at the beginning of the seventeenth century. The classic doctrine of the freedom of the seas (*mare liberum*) was devised by the Dutch scholar, lawyer and diplomat Hugo de Groot, more generally known as Grotius. Regarded as the founding father of modern international law, he advocated a narrow territorial sea under sovereign jurisdiction leaving an extensive high seas area, having the characteristics of an open-access commons. Grotius' view was to triumph over the more restrictive approaches of contemporary Portuguese and English jurists. The customary law of freedom of the seas was, over the ensuing three centuries, to accord not only with the interests of seafaring powers (notably Great Britain) but also with available technological capabilities. By the middle of the twentieth century the normative rights associated with the high seas commons were the 'four freedoms': to navigate, fish, lay cables and pipelines and to overfly. This long-standing regime for the high seas commons, allowing freedom of resource extraction on an essentially 'first come first served' basis, had already come under pressure through the application of industrial techniques and more sophisticated equipment to the harvesting of maritime resources. The introduction of explosive harpoons, whaling factory ships, and large deep-water stern trawlers began to reveal that what had seemed the limitless resources of the oceans were in fact finite. A dramatic demonstration was provided by the collapse of great whale stocks in the interwar period. It was also realized that previously productive deep-water fisheries required management if the 'tragedy' of overfishing was to be avoided. The response was the creation, in the first half of the twentieth century, of a number of regional common property resource (CPR) regimes known as fisheries commissions. In their attempts to manage stocks they abridged the absolute freedoms of the old high seas regime.

A combination of improved scientific understanding and new technologies then added a set of novel issues to considerations of the ocean commons. The most important economic prospect was the possibility, and by the 1960s the reality, of offshore oil drilling on the continental shelf. To this was added the discovery of apparently rich mineral deposits on the deep ocean floor in the form of manganese or polymetallic nodules. Whereas offshore drilling for oil and gas became commonplace, both the technology and economics of deep seabed mineral extraction remained problematic. Inputs of pollution also began to be recognized as a problem alongside the extraction of finite resources – indeed the two were often intimately connected. The emergent pollution issue received substantial public exposure through dramatic oil spills, the contamination of seafood and the collapse of some marine habitats. Much degradation was not associated with strictly maritime activities at all, but with the generally unrestrained use of the seas as a 'common sink' for the free disposal of land-sourced effluents. The regulation of the ocean commons thus ceased to be the simple matter it had been at the time of British maritime predominance, which now represented in Keohane & Nye's (1977: 88) inimitable phrase, 'a bygone era of fish and ships'.

The Law of the Sea

These developments encouraged national moves towards enclosure so as to garner the newly discovered mineral resources and to control and protect fish stocks. The US Truman Proclamation, which laid claim to the natural resources of the continental shelf but still referred to the waters superadjacent to the shelf as the high seas, led the way in 1945. It was soon rivalled by Peruvian and Chilean claims to resource jurisdiction and even sovereign rights extending 200 miles from their coasts. In response, there were two attempts to reform and codify the relevant international law: the UN Conferences on the Law of the Sea held in 1958 and 1960 (UNCLOS I and II). Keohane & Nye (1977: 92) see these events in terms of the erosion of the old 'freedom of the seas' regime and its replacement with, first a strong, and then a weak 'quasi regime'. The attempt to extend and modify the existing customary law was clearly insufficient and resulted in pressure, during the late 1960s, at the UN General Assembly and Seabed Committee for thoroughgoing reform.[1] The United States and the Soviet Union, as great naval powers, strove to regulate and check the rush towards enclosure of the high seas commons. They and other maritime powers were mainly concerned with the maintenance of navigation rights. *Ad hoc* extensions of national jurisdiction 'contained the threat that up to 114 key straits would be overlapped by territorial seas' (Schmidt, 1989: 22). Although the developed maritime states would have preferred a limited negotiation on navigation rights and the new limits of territorial seas, the very different requirements of the Group of 77, stressing national economic sovereignty and issues of North–South equity, could not be ignored. The politically necessary conjunction of the two provided the essential basis for the Third Law of the Sea Conference (UNCLOS III) and its extraordinarily broad and inclusive agenda.

This was to be by any standards a diplomatic epic, the like of which may (many commentators and participants fervently hope) never be seen again. It lasted a decade from its inception in 1973; involved some 93 weeks of formal negotiation (and many more devoted to preliminaries and informal meetings); and produced a Convention comprising 320 articles and 9 annexes ranging across the whole spectrum of maritime issues. From an 'ecoholistic' standpoint the complaint is often made, and will be made in subsequent chapters, that issue areas and hence regime arrangements are arbitrarily defined without reference to the boundaries of natural systems. It would be difficult to make this criticism of UNCLOS III. Here was a real attempt to codify an overarching regime for the oceans. UNCLOS III may, instead, be subject to the contrary criticism that the portmanteau inclusion of issues, apparently only related by their affinity to salt water, led to near impossible levels of negotiating complexity and excessive scope for linkage politics. In the end it was linkage between the controversial seabed provisions and the rest of the Convention which was to provoke the refusal of an incoming US administration to endorse the fruit of a decade of diplomatic handiwork. Yet, as Ogley (1984: 239) notes, the Conference was 'unfortunate in having to encounter one of the most dramatic reversals in American political history'.

The connection of issues as diverse (although often ecologically connected) as rights of navigation through archipelagos, the sustainable harvesting of fish stocks, and the mining of the seabed, was sanctioned by the legal and institutional history of the Law of the Sea but also by the political realities of the UN General Assembly. The concerns of the maritime states have already been mentioned. They were confronted by a number of developing countries, particularly in Latin America, which were ardent advocates of 200 mile limits. Such demands were in line with the developing Group of 77 doctrine of economic sovereignty but tended to cut across the wider concern with justice and equity in North–South relations and the campaign of substantial numbers of Group of 77 states for the establishment of a more equitable 'common heritage' regime for the oceans. The latter reflected Arvid Pardo's speech of 1967, which, as we have already seen, had ramifications far beyond the Law of the Sea. It evidently did much to set the agenda for UNCLOS III, and resonated with the broader objectives of the New International Economic Order (which the Pardo speech pre-dated by several years). For members of the G77 there remained a central political paradox. The establishment of 200 mile exclusive economic zones (EEZs) for a privileged group of coastal LDCs, drastically reduced the commons area open to incorporation within a common heritage regime.

The range and complexity of the negotiations and of the interests involved led to a complicated set of alignments that shifted across the various linked issues. New groupings, such as that for 'geographically disadvantaged and landlocked states', were created alongside the '200 milers' and the developed maritime states. On specific issues cross-cutting coalitions would form. It was no small achievement that the negotiations were concluded at all, especially as it had been initially agreed that the only basis upon which progress would be made was that of an eventual 'package deal' in which nothing would be finally decided until everything was decided. The compromise nature of any agreement and in particular the status of the seabed arrangements as a quid pro quo for other parts of the Convention was well understood.

The outcome, as represented in the 1982 Montego Bay Law of the Sea (LoS) Convention, was to 'enclose' large tracts of sea and seabed which had previously been *res nullius* or subject, in the case of fisheries, to a CPR regime. Coastal states were allowed a 200 mile EEZ along with economic rights to the continental shelf extending, in some special cases, out to 350 miles.[2] Accordingly the, now much reduced, ocean commons were designated as the high seas and what was defined as the 'Area' meaning the 'seabed and the ocean floor and subsoil thereof, beyond the limits of national jurisdiction' (Art. 1(1)). The really novel feature of the LoS Convention was the attempt, alongside the enclosure of EEZs, to establish a common heritage regime for the mining of the deep seabed, although Pardo (1984) himself was to express his disappointment with much of it. Despite being negotiated by both the Ford and Carter administrations as part of a package it was to prove the 'sticking point' for the Reagan administration which, along with Britain and the then West Germany, refused either to sign or ratify the Convention. For 11 years the Convention failed to acquire the necessary ratifications (60) until 16 November 1993 when Guyana became the 60th state to deposit an instrument of ratification. Entry into Force was to occur one year subsequent to that date but the difficulty remained that the United States, Canada, the states of the European Union and Japan were conspicuously absent from the list of those acceding and ratifying. From 1990 the Secretary General held a series of 'consultations' designed to promote 'universal participation' in the LoS Convention. As will be discussed later in this chapter, what this in effect meant was the re-negotiation of Part XI, containing the controversial seabed regime, so as to make it acceptable to the major developed world states.[3] Despite all this, the EEZs and other provisions of the Convention had already gained acceptance as international customary law. In fact, widespread adoption of national EEZs actually occurred in 1976–77 before the completion of the Convention.

In one sense it would be possible to consider the provisions of the LoS Convention as establishing a single regime for the oceans. The legal users of the term would have no difficulty in defining it as such and the negotiators at UNCLOS III quite consciously aggregated a number of issue areas together. However, the conception of regime in International Relations or International Political Economy is issue area specific. Thus we may identify three main global maritime issue areas. The first, the whaling issue area, concerns a truly global common pool resource. Its regime, based upon on a 1946 agreement, pre-dates all three UN Law of the Sea Conventions.

The second issue area identified is much more inchoate and addresses a single 'common sink' problem with many sources – pollution of the oceans. It might be a matter of debate whether one or several partial issue areas and regimes exist in this area. Both whaling and pollution regimes have a relationship to the 1982 Convention, which may be seen as providing an authoritative set of principles and norms relating to both the harvesting of maritime resources and the prevention of environmental degradation.[4] The third issue area, directly dealt with in the Convention, is the deep seabed. Here was the first attempt to design a common heritage regime for a common pool (mineral) resource.

Whaling

At the international level the attempt to bring open access fisheries under some from of CPR regime has long posed problems. Various forms of governance have been devised – notably the fisheries commissions. None of these had global scope, all being related to a particular regional fishery. Examples include the Indo Pacific Fishery Commission, the Inter American Tuna Commission and the North East and North West Atlantic Fishery Commissions.[5] They performed collective scientific and management functions, establishing and then attempting to draft and enforce rules for 'maximum sustainable yield'. Establishing the latter was a scientifically difficult task in itself and required a sophisticated understanding of fish population dynamics. Effective rule-making ran up against the standard difficulties of administering a scarce and depletable common pool resource at the international level. The record was not one of unalloyed success and was, according to one authority, 'inversely related to the number of member nations' (Driver, 1980: 44). The significance of the fisheries commissions has been diminished by the proclamation of EEZs (covering 35% of the world's sea area but 90% of living resources) which placed most of the key fishing grounds, that had once had the status of commons, within the economic control of particular state authorities. This has not solved the problem of overfishing. Insatiable demand for fish, overcapitalized fleets and ineffectual national regulation have led to a situation where, in 1994, the FAO estimated that nine of the world's 17 main fishing grounds were in potentially catastrophic decline. The problem is particularly serious in the relatively unpoliced waters of the South now fished by developed world fleets which have exhausted their own fishing grounds.[6]

With the exception of the Antarctic CCAMLR (see p. 75) the primary global arrangement for a common marine property resource covers whaling. There has also been concern for the conservation of 'straddling stocks' such as the extravagantly named Southeast Pacific Jumbo Flying Squid (which ranges from California to Chile). A relevant UN Convention incorporating the precautionary principle and cooperative conservation management between coastal and fishing states was negotiated during the early 1990s and concluded in 1995.[7]

Whales, or more precisely the 79 species of cetaceans, are often highly migratory, and for a long period were the basis of a significant industry producing whale oil, whalebone (from the baleen plates which the majority of whales employ to sieve their food) and a wide variety of other products including meat, still regarded in Japan as a particular delicacy.[8] The long history of the whaling industry displays the cycles of boom and bust characteristic of unregulated fisheries, only exacerbated by the lengthy period over which whales grow to maturity and the peculiar voracity of the hunters. To this was added the extra 'efficiency' of explosive-tipped harpoons and the stern slipway factory ships of the interwar period. When first American and then North Atlantic stocks were exhausted, attention turned to the rich hunting grounds of Antarctic waters. As one species (for example the great blue whale) neared extinction, other smaller or less accessible whales were hunted until by the 1980s the only species left in commercially exploitable quantities was the relatively small (10 m long) minke whale. The result of all this was an open access 'commons tragedy' on a daunting scale in which an over-capitalized whaling industry took its short-term profits and then, in the main, disappeared along with most of its quarry. An even earlier collapse was only averted by the First and Second World Wars, which gave rise to an incidental cessation of whaling during which stocks were able to recover somewhat. In the early 1930s there were attempts by whaling companies to erect a regime of mutual production restraint. But this essentially cartel-like activity appears to have been motivated by the need to maintain and support oil prices in a depressed market rather than any concern with long-term sustainable management. Awareness of the latter was manifested in international agreements signed in 1931 and 1938, but a real attempt at a CPR regime was only to emerge with the International Convention for the Regulation of Whaling (ICRW) of 1946.

Norms and Principles

The 1946 Convention recognizes in its preamble the open-access resource problem and the essential need to 'protect all species of whales from further over-fishing'. The concern of its signatories, the whaling nations, was nonetheless most definitely with the managed exploitation and the 'safeguarding for future generations of the great natural resources represented by the whale stocks'. At that time post-war shortages led to an emphasis on the production of whale oil. In the view of the UK official participating in the 1945 Conference that negotiated the ICRW, this was the overriding concern. It would have been a 'tragedy if any international machinery were to get in the way of increased production'.[9] The belief was that whale stocks and indeed the catch might be increased if the industry was to be properly regulated and that this might be achieved without 'endangering these natural resources'. The common interest was to achieve 'the optimum level of whale stocks', confining whaling to those species 'best able to sustain exploitation' (ICRW Preamble). The problem was that the 'optimum' was not defined and was certainly not interpreted in terms of the maximum sustainable yield concept developed for other fishery resources. Neither was there any sense that whales should be recognized as a highly mobile 'common heritage' resource in which everyone had a legitimate interest. The regime may have urged conservation but it was still very much an arrangement between those with an interest in commercial exploitation.

For many years the ICRW remained largely ineffective. Whale stocks declined along with the whaling industry. Countries such as Britain and the United States effectively gave up commercial whaling (the last British ship was sold to Japan in 1963) but remained within the Convention, where they were joined by a number of new members. The collapse of the industry, the changing composition of the ICRW membership and, above all, the rise of widespread and well articulated public and NGO concern about whales, led to a dramatic shift away from the original principles of a regime for sustainable exploitation towards one of outright prohibition. The change had been signalled by the 1972 Stockholm Environment Conference which passed a unanimous resolution calling for a moratorium on whaling, but it took the International Whaling Commission (IWC) until 1985/86 to implement such a measure. Many supporters of a moratorium had come to regard it as morally wrong to butcher highly developed mammals sharing both some of the characteristics and indeed the rights of *Homo sapiens*. For environmentalists the campaign transcended the specifics of the prohibition of whaling:

> Over the past two decades, the battle to save the whales and dolphins has symbolised the dawning of a more environmentally conscious age with the determination to fight against those destroying the planet for short-term profit. The fight for these animals is a fight for the soul of environmentalism. Today this fight rages harder than ever.[10]

There is little doubt that the anti-whaling campaign was of pre-eminent importance for a range of well known NGOs, Greenpeace, the WWF, the International Fund for Animal Welfare and the Environmental Investigation Agency, to name but a few. They were able to mobilize a widespread popular distaste for whaling amongst Western publics. Furthermore, they managed to install themselves both as observers and as delegation members at IWC meetings. This activity had an effect upon a number of leading signatories of the ICRW who no longer had a direct commercial interest in whaling. It was responsible for a shift in national policies concerning the fundamental purposes of the Convention. The outcome was the 1986 IWC moratorium on commercial whaling and subsequent attempts to redefine regime principles such that 'optimum utilization' would involve so called 'non-consumptive' activities such as 'whale watching'. The latter is now estimated to constitute a $500 million a year industry enjoyed by around six million people.[11]

The prohibitionist view seemed in the mid-1980s to have majority support in the IWC, but was bitterly contested by the few remaining whaling nations, notably Norway, Japan and Iceland. No less than Gro Harlem Brundtland joined the contest in support of a resumption of catches by the whaling communities of northern Norway:

> We have to base resource management on science and knowledge, not on myths that some specifically designated animals are different and should not be hunted regardless of the ecological justification for doing so. International cooperation is in danger if this kind of selective animal welfare consideration is allowed to dictate resource policies.[12]

The controversy between the advocates of animal rights and the sustainable hunting of whales contained a heady mix of domestic politics and ethnocentrism. While for countries like Britain, the United States and France, opposition to commercial whaling might be a relatively cost-free way of establishing 'green' credentials, for the Norwegians and Japanese support for resumption constitutes action in defence of traditional communities and national rights. They assert what are, in their view, the proper unsentimental principles of the regime involving the sustainable harvesting of what is, in the end, just another natural resource. This central conflict of principle has continued to bedevil the operation of the whaling regime.[13]

Organization and Procedures

The ICRW set up the IWC which meets annually and is composed of a commissioner from each member state. There were 15 original signatories, all involved in commercial whaling. In 1999 membership had expanded to 40, of which only two, Norway and Japan, had whaling industries (Iceland 'withdrew' in 1991). The composition of the IWC is important because there is a majority voting rule and a requirement for three-quarters of the members present and voting to approve changes to the 'Schedule' containing the operative rules of the regime (Art. III). Nor is there any 'activity' criterion (as may be found in the closely related regime for Antarctic Marine Living Resources) making practical or scientific involvement with whale industries, whether hunting or watching, the basis of membership. Thus, landlocked Switzerland was able to become a member in 1980. The critical alteration of the regime through the introduction of a moratorium on commercial whaling, which finally achieved the necessary majority at the 1982 Brighton IWC, was based upon quite dramatic changes in membership. Thirty-seven members participated in the key vote, including eight new members, none of which had whaling industries. This was sufficient to carry the day for the moratorium 25 for, 7 against and 5 abstentions.[14] However, 'reservations' from collective decisions are allowed and it is on this basis that the Norwegians have asserted their right to continue whaling. Iceland pursued a different strategy in leaving the IWC and setting up, in 1992, an alternative regime in the shape of the North Atlantic Marine Mammals Commission (NAMMCO). Involving the Faroe Islands, Greenland, Iceland and Norway, NAMMCO, which has its own scientific and management committees and a wider remit than the IWC (extending to seals and dolphins), represents a distinct threat to the existing regime. As Stoett (1997: 81) notes: 'No non-whaling members belong thus directly challenging the notion that the commons should be managed by as many states as possible'.

The inclusive nature of the existing regime and the organizational features of the IWC provide ample scope for what can only be described as influence peddling. Japanese activity in providing financial assistance to various Caribbean states is well known. At the 1993 Kyoto IWC meeting it was reported that Japan had enlisted the support of four Caribbean nations, Dominica, Grenada, St Lucia and St Vincent. Apparently 'Japanese aid to these countries' fishing industries had been tied to the payment of affiliation fees to the IWC and the appearance of delegates at the meeting'.[15] However, NGOs have also 'sponsored' national delegations with a British conservation group providing financial assistance for Seychelles participation (Stoett, 1997: 66). Although the financial scale of the IWC is small (an annual budget of £1.3 million in 1998), it has maintained a permanent secretariat, based first in London and then Cambridge, UK. Its annual Commission meetings are, however, peripatetic.

The IWC is not the only organization involved with the whaling regime although its position is central. As with many environmental regimes there is a significant trade dimension. In this instance the connection is with the Convention on Trade in Endangered Species (CITES). Whales are included in Appendix I of CITES, which lists those species upon which Parties must impose an absolute trade ban. At the 1997 Meeting of the Parties in Harare the conflict within the IWC spilled over into CITES as Norway and Japan sought to weaken the whaling moratorium by proposing the 'downlisting' of certain whales from Appendix I.[16] If the move had been successful (it fell short of the two-thirds majority required), it would not only have legalized trade in whale products but would have had clearly detrimental implications for the authority of the IWC.

Rules, Information and Enforcement

The operative rules are contained in the schedule to the ICRW and were originally envisaged (in Art. V) as covering: protected species, open and closed seasons and areas (including sanctuaries), limits on the sizes of whales to be caught, and a total allowable catch (TAC) for each season. There were also to be specifications as to gear and a requirement for members to collect catch and other data. All regulations were to be based on scientific evidence and were not to involve specific quotas for individual whalers or groups of whalers. A major loophole in any future regulation was presaged by Art. VIII which allowed the taking of whales for scientific purposes notwithstanding the rest of the agreement, subject only to such restrictions and conditions as the contracting government 'thinks fit'. This was to be exploited on a huge scale by the USSR and to a lesser extent by the Japanese. Another loophole exists in the recognition of subsistence 'aboriginal' whaling where communities are allowed to continue their tradition of hunting whales on a limited scale. Finally, the regime was designed to cover only great whales, thus excluding a large number of smaller whales (the majority of the 79 species of cetaceans). The need to extend the protection afforded by the ICRW became a key issue in the 1980s.

The major innovation in the 1946 ICRW was the agreement to fix an overall catch limit or quota. This became the most highly publicized annual activity of the Commission and one which, up until 1970, involved a visible failure to fulfil the original purposes of the ICRW. The quota was always set too high and in terms of the discredited blue whale unit (BWU). This took no account of different species and measured catches in terms of the assumed yield of whale oil using the blue whale as a yardstick. Thus one blue whale was deemed equivalent to several smaller whales and the system provided no incentives to discriminate in ways which would have conserved stocks.[17] In terms of comparable fisheries agreements these rules were primitive indeed and the IWC lacked both the essential scientific understanding and data and the specific policy instruments to tackle the problem of the progressive collapse of whale stocks. Accordingly 'quotas were never sensitive to the particular stocks, nor were they set at levels that could ensure sustainability and wasteful competition and depletion was not avoided' (Sand, 1992: 258). Whalers were actually unable to catch the allotted quota and the absence of effective restrictions meant that the catch of whales doubled from 31 000 in 1951 to 66 000 in 1962, with the blue whale facing extinction. 'In fact, more whales were killed under the new regime than had been killed when whaling was unregulated' (Porter & Brown, 1991: 79). The shift towards prohibitionist norms led by the United States and underlined by the 1972 UNCHE's call for a 10 year moratorium led first, in 1975, to the introduction of New Management Procedure rules controlling catches in the interest of sustaining whale stocks and then to the 1982 IWC decision for a moratorium or zero quota from the 1985/86 season.[18] The official rationale for the moratorium was to allow the development of a revised management procedure (RMP) by the Scientific Committee to allow a scientifically justifiable catch. Yet, although the plan was approved by the IWC in 1994, anti-whaling nations have been able to obstruct its implementation.

The other form of regulation employed by the regime has been the designation of sanctuaries, areas off limits to commercial whaling originally on a temporary basis. In 1979 an Indian Ocean Sanctuary was designated, renewable in 2002. At the Mexico IWC meeting of May 1994 a French proposal for an Antarctic sanctuary, south of 40 degrees, was accepted against the isolated opposition of Japan. Because the sanctuaries apply to commercial whaling activities they must be seen in relation to the ending of the moratorium and the adoption of the RMP which would allow North Atlantic and Pacific catches of minke whales. It would have the effect of banning Japanese commercial whaling in Antarctic waters while allowing Norwegian whaling in the North Atlantic.[19] As of 1999, however, the majority resistance in the IWC to any commercial whaling has meant that the RMP remains unimplemented.

Under the Convention both observation and enforcement are in the hands of contracting governments (Art. IX). It took until 1959 for the ICW to agree an International Observation System and until 1972 for it to be implemented (the plan was to place two national observers and one international observer aboard each ship). Before that there had been no systematic observation of the extent of compliance with the rules and participants acted on the assumption that their competitors would flout them. A depressingly ruthless exploitation of remaining great whale stocks was the almost inevitable consequence. The observer system was not centrally administered and constituted a set of reciprocal arrangements between ships whaling in the same region (Rose & Rowland, 1993). It appears to have worked reasonably well except in detecting the complete disregard of the rules by Soviet ships in the period. It may be doubted whether any scheme would have been effective in this case for no less than the KGB was reportedly involved in the encryption of communications between Soviet whalers along with other forms of elaborate subterfuge.[20]

Following the 1985/86 moratorium reporting requirements have been reduced and only involve the six members who still take whales for the allowed 'subsistence' and 'scientific purposes'. The ICRW allows members to opt out of rules in the Schedule by the simple expedient of filing an objection within 90 days (Art. V(3)). Four nations used this procedure against the moratorium but only Norway openly resumed commercial whaling. Although there are no enforcement provisions in the ICRW, national measures associated with the

IWC rules have been unilaterally threatened by the United States under the stimulus of conservation groups influential in the Congress.[21] The legislation exists to impose import bans and the revocation of rights to fish within the EEZ. More impressive are the actual or threatened boycotts of Japanese, Icelandic and Norwegian produce organized by supporters of the environmental NGOs. On occasion firms have been persuaded to take action: Burger King, for example, cancelled a $5 million dollar order for Norwegian fish.[22]

The ICRW stresses the need to take action only on the basis of reliable scientific evidence; however, it did not create a dedicated independent body to generate and evaluate scientific data. Instead, Art. IX speaks of collaboration with other agencies and governments in the encouragement of research and the collection, analysis, evaluation and dissemination of findings on the methods of 'maintaining and increasing the populations of whale stocks'. Under Art. III(4) a Scientific Committee composed of members, designated by national Commissioners' was set up and occupies a crucial position because of the requirement (Art. V(2)b) that changes in the rules contained in the schedule should have a scientific basis. Over the years the Scientific Commission has increased in importance as new management procedures have been introduced but has relied heavily on the professional expertise of advisers from other international organizations (FAO, UNEP, IUCN) which have observer status. However, the Committee has been hampered, first, by being 'over-willing to temper its recommendations to conform to so-called "practicalities"' (Birnie, 1985: 261) not unconnected to the commercial interests of member states and, latterly, by an inability to make any recommendations at all on some issues because of a lack of both theoretical understanding and empirical data. There are major gaps in knowledge of whale migration patterns and indeed of the size of stocks. This is an area where the science has become heavily politicized and where absence of knowledge has provided excuses for inaction. Paradoxically, inaction in the whaling case has meant a failure to adopt the RMP and maintenance of the moratorium which fulfils the objectives of anti-whaling and non-consumptionist opinion. Whether the regime can survive such obstructionism in the longer term remains to be seen. In strictly ecological terms the major threat to cetacean populations probably no longer arises from 'harvesting' activities. Instead, rising levels of marine pollution and chemical contamination plus the depletion of fish stocks and stratospheric ozone loss pose a greater threat. Such problems are now being addressed as a matter of priority by the Scientific Committee's Standing Working Group on Environmental Change.[23]

Marine Pollution

The oceans provide the greatest of all common sinks for the wastes of human activity. Jacques Cousteau said that there is in fact only one pollution 'because every single thing, every chemical whether in the air or on land will end up in the ocean'.[24] Up until the middle of the twentieth century pollutants might well be regarded literally as the proverbial 'drop in the ocean'. However, the great increase in the scale and potential damage of pollution was demonstrated by such mishaps as the *Torrey Canyon* wreck of 1967 which spilled crude oil into the English Channel or the fate of the villagers of Minimata in Japan, where mercury infiltrated the food chain through contaminated fish. These two incidents highlight the sources of marine pollution. First, there is pollution from ships either by accident or as a consequence of day-to-day operations. This accounts for 34% of the 3.2 million tonnes of oil deposited annually in the oceans (Tolba & El-Kholy, 1992: 122). Then there is the deliberate dumping of industrial and other wastes at sea. Some of these may have lethal effects, not only for maritime ecosystems, but on human consumers as well. Land-based pollution through sewage outflows, the discharge of industrial effluent or the 'run-off' of fertilizers poses the pollution problem of the greatest magnitude for coastal regions in particular. The estimate in the *Agenda 21* document is that land-based sources contribute 70% of marine pollutants while maritime transport and dumping account for 10% each.[25] Finally there are the actual and potential effects of offshore oil drilling and seabed mineral extraction.

The oceans can be treated as a global commons but effective action on enclosed seas and coastal regions may more suitably be based on the concept of a local or regional commons. Most of the pollution originates from territories or EEZs under national jurisdiction and the special vulnerability of enclosed seas such as the Mediterranean and the Persian Gulf may require much more stringent controls than would generally be required. Following the logic of 'transaction costs' arguments about cooperation, such controls may also be easier to arrive at in negotiation between a relatively small number of directly interested parties. These considerations have been reinforced by the introduction of EEZs and the enclosure of the main fishing grounds and shallow coastal areas most likely to be damaged by pollutants.

A global maritime pollution regime has only been developed since the 1972 Stockholm Conference. This provided the stimulus for two global agreements, both dealing with pollution from ships: first, the 1972 London Convention, which treats the dumping of industrial waste. Second, the MARPOL (more formally the International Convention for the Prevention of Pollution from Ships, first signed in 1973 and amended in 1978), which regulates discharges of oil and other substances. Institutions relating to the more significant land-based sources of pollution are much less well developed. UNEP's 1985 'Montreal Guidelines for the Protection of the Marine Environment Against Pollution from Land-Based Sources' are, as their name suggests, 'soft law'. They impose no obligations and exist to structure and encourage the negotiation of bilateral and regional anti-pollution agreements. As was demonstrated in the drafting of the environmental provisions at UNCLOS III, states, and particularly developing states, have been far less willing to enter into obligations relating to their land-based activities than relating to ship-based pollution. The political arguments were not just between coastal and maritime states, but also involved the special case that could be made out by developing states regarding their land-based pollution and the assertion that they could not be expected to adhere to the same kind of emission standards as the developed world. In Art. 207 pollution control is weighed against what may explicitly be the more pressing requirements of economic development. Thus control measures must take into account: 'characteristic regional features, the economic capacity of developing states and their needs for economic development' (this formulation is repeated in Arts 4 and 5 of the Montreal Guidelines).

An enormous amount of effort has already been devoted to the construction of a diverse institutional and legal architecture covering aspects of maritime pollution: from global agreements to liability and compensation schemes, safety conventions and arrangements for cooperation in the event of pollution emergencies. As of 1988 there were 'over 70 multilateral instruments of one kind or another organized in 40 "clusters" of related treaty arrangements' (Johnston, 1988: 199–200). UNEP's Regional Seas Programme alone comprises 11 regional arrangements and there are three additional schemes for the North Sea, Nordic Sea and Baltic.

Arguably this is another area in which, despite the overarching nature of the provisions of the 1982 LoS Convention and the ecological understanding of the integrated nature of the problem, a number of distinct issue areas exist and regimes are fragmented. There is, as yet, no institutional structure to handle land-based pollution and thus the need for the 'creation of a global mechanism to co-ordinate the protection of the marine environment from all sources' was a key recommendation of the parties to the London Convention to the 1992 UNCED.[26] The point was duly noted in *Agenda 21* (Ch. 17: B). Subsequent to this there have been encouraging moves towards a more integrated approach to marine pollution. The 1996 Protocol to the London Convention (which effectively replaces it) essentially rules out sea-based waste disposal in favour of sustainable land-based solutions. More significantly, it shifts the domain of the regime landwards through the encouragement of collaboration with local and national agencies concerned with pollution control from both point and non-point land-based sources.[27]

Norms and Principles

For centuries the right to pollute was an implicit freedom of the high seas. Given the institutional diversity of the pollution regime or regimes it might seem difficult to isolate the way in which general norms and principles have evolved. In this as in other areas UNCLOS III represented a systematic attempt to codify such changes. In terms of principle, there was a belief that regulation had in future to be based, not on specific economic and indeed human interests which ought to receive protection, but on the welfare of the whole marine environment. Article 194 speaks of 'pollution of the marine environment from any source' and measures necessary to 'protect and preserve rare and fragile ecosystems as well as the habitat of depleted, threatened or endangered species and other forms of marine life'. The norm is that states have 'the obligation to protect and preserve the marine environment' (Art. 192). Echoing the famous Stockholm Principle 21, they also have the 'sovereign right to exploit their natural resources pursuant to their environmental policies and in accordance with their duty to protect and preserve the marine environment' (Art. 193). This may present a substantial obstacle to the creation of an integrated regime for the control of land-based maritime pollution. Yet, norms and principles that are elsewhere frequently recommended for pollution control regimes are absent. *Agenda 21* urges a 'precautionary and anticipatory rather than a reactive approach to prevent the degradation of the marine environment' involving such things as mandatory impact assessment, standards for the handling of hazardous substances and clean production techniques along with a comprehensive approach to damaging impacts (Ch.17: B). Dumping of waste at sea was not regarded as inadmissible *per se* by the London Convention. Instead Annex 1 referred to a prohibition on particular hazardous substances on grounds of 'toxicity, persistence, bioaccumulation and the likelihood of significant widespread environmental exposure'. The approach has now been modified such that the 13th Consultative Meeting of the parties to the London Convention agreed that the dumping of all industrial wastes (other than of those of an inert nature) should be terminated by the end of 1995 (Tolba & El-Kholy, 1992: 124). At the next Consultative Meeting in 1991, the precautionary principle was explicitly adopted. This entailed a 'shift of emphasis from controlled dumping, based on an assumption of the assimilative capacity of the oceans, to approaches based on precaution and prevention'.[28]

Organization and Procedures

The existing agreements of the maritime pollution regime(s) share a common organizational basis in the International Maritime Organization (IMO). It acts as a secretariat, convening meetings and handling information generated by the reporting requirements of the two global conventions.[29] The supreme decision-making body of the London Convention is the Consultative Meeting of the Parties held at maximum intervals of two years. The MARPOL has no such body but has developed the Convention through the Marine Environment Protection Committee (MEPC) of IMO. The MEPC also serves to coordinate with a range of related IMO and UNEP based conventions. Neither Convention created a dedicated source of scientific advice but both are dependent in fulfilling their functions on scientific and technical assistance (to determine, for example, the ecosystemic implications of the dumping of particular substances). Here the regime(s) are supported by a complex and long-established international network devoted to the study of oceanography and maritime resources. This dates back at least to 1902 with the foundation of the International Council for the Exploration of the Seas (ICES). At present the key advisory body on matters concerning the impact of maritime pollutants and their control and abatement is the Joint Group of Experts on Scientific Aspects of Marine Pollution, more pithily described as GESAMP. This body is formally an interagency programme sponsored by seven UN agencies including UNEP, IMO and IAEA.

Rules, Information and Enforcement

The first substantive attempt to devise international rules relating to oil pollution from tankers occurred in 1954 with the London Convention. Over the years it set standards for tanker construction and attempted to impose limitations on oil discharges. The 1973/1978 MARPOL Convention built on this approach and increased the number of areas (enclosed seas and fragile maritime ecosystems) where no discharge of oil was permissable. MARPOL also extended the rules to cover other types of ship-based pollution resulting from the carriage of hazardous substances and the throwing overboard of garbage. The Convention itself is rather brief, the real meat of the regulations being contained in the various annexes. These contain an extraordinary amount of dense technical detail specifying, to give one example, standards for the dimensions of flanges for discharge connections (Annex I Regulation 19). This is to 'enable pipes of reception facilities to be connected with the ship's discharge pipeline for residues from machinery bilges'. Concentration on technical standards for the construction and equipment of tankers has led to greater 'enforceability' in that such provisions are easier to monitor than regulations on discharges (Mitchell, 1993). In the 1992 Protocol there was a new emphasis on the prevention of pollution through accidents. This was achieved by specifying that both new *and existing* tankers have double bottoms and wing tanks extending the full depth of the ship's side.[30]

The London (Dumping) Convention provides a 'much more stringent regime than is found in most regional controls on land based pollution' and 'is widely regarded as one of the most successful regulatory treaties' (Birnie & Boyle, 1992: 321). The regulatory system operates on the basis of permits. No substances may be dumped at sea without the authority of a permit to be issued by the national authorities within whose territory the waste is loaded.

The key detail is in three annexes to the Convention. Annex I lists substances that may not be dumped at all such as mercury, cadmium, persistent plastics and high-level radioactive waste. Annexes II and III list substances that may be dumped under a specific or general permit respectively. Starting from an original presumption that some dumping was acceptable the rules covering specific substances have been progressively tightened to reflect the precautionary principle. In 1993, what had been a voluntary moratorium on the dumping of low-level radioactive waste, was translated into a formal ban. The effect of the 1996 Protocol is to change fundamentally the list system described above. Thus, rather than prohibiting the dumping of certain listed materials, the Parties to the Protocol are obligated to prohibit the dumping of any waste or other matter not listed in Annex I. This 'reverse list' system, thus requires a permit for the dumping of previously non-prohibited substances.[31]

Enforcement of MARPOL, as with other maritime regulations, is in the hands of the 'flag state', i.e. the state registering a particular ship. A total of 85% of the world's merchant tonnage is registered with states signatory to MARPOL. Flag states are required to inspect vessels and issue them with international oil pollution prevention certificates (Art. 5) but they are additionally subject to inspection by the 'port states' to which they sail, which may detain them for repairs (Art. 6). All parties also undertake to report violations the flag and or port states concerned. The certification and inspection system is applied somewhat patchily with a well organized collective system of inspection being operative in Western Europe under the 1982 Paris Memorandum of Understanding. MARPOL has no systematic compliance procedure relating to the behaviour of its members. Instead there are mandatory reporting requirements under Art. 11 whereby national reports filed with IMO are circulated to the parties. This system of 'horizontal enforcement' is vitiated by the fact that only around 30% of signatories bother to report and even then the quality of information is frequently inadequate (Sand, 1992: 162). Enforcement is very variable, with a few developed states such as the US, Japan and Canada, being relatively zealous whereas others make very little attempt (Mitchell, 1993: 230–231).

The London Convention has a stable membership of around 70 states. It relies exclusively on enforcement by the states within whose jurisdiction waste is loaded for dumping. They are required to assess the likely environmental impact and issue permits to dump. They must, of course, ensure the prohibition of dumping of various hazardous substances as required under the Convention. This situation applies to the high seas. Within 200 mile EEZs the coastal state has jurisdiction which may be modified by membership of various regional agreements (e.g. the Oslo Convention for the North East Atlantic and North Sea) that are generally tighter than the provisions of the London Convention, and often involve international inspection. This is conspicuously absent from the London Convention something that may be regarded as a major failing of the high seas dumping regime. Compliance provisions are similar to those for MARPOL in that parties are required to notify the IMO of their issue of permits and to report their compliance monitoring and impact assessment activities. Once again the fulfilment of such obligations is far from comprehensive.[32]

The verdict on the effectiveness of the maritime pollution regime(s) must be mixed and to a certain extent open. There is a major and continuing problem with the failure to produce effective regulation of land-based pollution. On the other hand evidence from GESAMP and from IMO commissioned studies shows that there has been a significant decline in oil pollution from ships during the 1980s. Although partly attributable, no doubt, to recession and changing patterns of energy use, the conclusion is that MARPOL has had 'a substantial positive impact'.[33] The London Convention can point to a similar record with a reduction in dumping of industrial waste from 17 million tonnes in 1979 to 6 million tonnes in 1987 (Sand, 1992: 154). The 1993 ban on 'radwaste' dumping is also significant. It marks an application of the precautionary principle and a victory for a long campaign of direct and other action, principally waged by Greenpeace. The ban reflected a shift of position by a number of major industrialized parties to the London Convention, including the UK and US, who had previously been advocates of the right to dump low-level waste from their nuclear, medical and military activities. Similarly, the 1996 Protocol represents a decisive shift away from the principle of regulated dumping to one of absolute prohibition. Taken together, these alterations suggest that regime change is occurring.

The Deep Seabed

Part XI of the 1982 Law of the Sea Convention, which contains the blueprint for mining of the deep ocean floor, represents the most ambitious ever attempt to create a global common property resource regime. In its attempt to embody concepts of equity and community interest into the arrangements for the exploitation of a commons resource it was intended to provide a practical application of 'common heritage of mankind' principles. Arvid Pardo's salient advocacy in 1967 was bound up with a widespread belief that new and untold mineral riches lay scattered on the deep ocean floor in the form of manganese (or more correctly polymetallic) nodules. The main metals of commercial interest known to be found in such nodules are nickel, copper and cobalt. Initial estimates (by the American geologist John Mero who single-handedly rediscovered the nodules and whose figures were used by Pardo) ran to billions of tonnes of recoverable metal reserves. Five multinational deep sea mining consortia were formed alongside an equal number of state concerns. As the negotiations proceeded through the 1970s estimates and enthusiasm began to wane as the costs and technological problems of dredging the ocean floor were better understood and metal market prices softened. It became clear that the main potential site for realistic mining was the Clipperton–Clarion fracture zone in the Pacific and that deep sea mining was not yet a viable economic prospect.[34] Pardo himself (1984) was to write that the seabed negotiations had been based on some false premises, regarding deep sea nodules as the only form of submerged mineral wealth (there were other types of rich deposit within the boundaries of national economic sovereignty) and the necessity that intending miners would treat with the organizations set up in the 1982 Convention. Nonetheless, given the rate at which land-based reserves are being extracted, it would be foolhardy indeed to claim that seabed mining beyond national jurisdiction will not become a reality.

Norms and Principles

The principles of the regime rest upon the assertion that the seabed beyond national jurisdiction, known as the 'Area' and its resources are the 'common heritage of mankind' (Art. 136). This would include a commitment to peaceful use referred to in Article 144 and, to an extent, already embodied in a 1971 Treaty which forbids the militarily unlikely activity of emplacing weapons of mass destruction on the seabed.[35] However, the entire focus of Part XI of the LoS Convention dealing with the seabed is on the exploitation of its mineral resources. Article 140 states that activities in the Area shall be:

> carried out for the benefit of mankind as a whole, irrespective of the geographical location of States, whether coastal or land-locked, and taking into particular consideration the interests and needs of developing states.

This is coupled with the norm that there shall be 'equitable sharing of financial and other benefits' and an obligation placed upon those mining the seabed to transfer the relevant technology to an international body (the Enterprise) and to developing countries (Arts 140(2) and 144).

In order to grasp the full import of what was proposed for the seabed, it is necessary to appreciate that although some form of regime would be required to ensure orderly mining and, above all, to provide investors with assurance as to the security of title to claims, the arrangements finally agreed were revolutionary in the extent to which they took the wider interests of the international community into account alongside those of miners.[36] What was finally achieved was a compromise between these interests and the two sets of opposed principles. These comprised, on the one hand, liberal ideas concerning access and the operation of the market and, on the other, those associated with 'common heritage' involving a more authoritative and equitable allocation of resources. In the protracted negotiations at UNCLOS III the tension between such principles underlay the conflict between the developed nations, led by the United States, and the Group of 77 coalition. As the representatives of potential mining entrepreneurs the former pressed for a regime (in some ways analogous to that for the radio frequency spectrum) where exploitation would be on the basis of 'first come first served' and where organized international-level involvement would be limited with the provision of some form of claims registry perhaps tempered by a compensatory element for the benefit of the international community including the landlocked. It was argued that only such a regime could provide the necessary economic incentives for entrepreneurs to invest capital in the risky and untried business of seabed mining. Such a regime would be most likely to generate revenues, some of which could be shared out amongst developing and geographically disadvantaged states. As with other such 'common pool' regimes, it would inevitably imply that, in the main, only developed countries could effectively utilize the common resource and that the technological gap between them and the less developed countries would remain.

The G77 position was to press for a centralized international authority which would mine the seabed and distribute the proceeds in a fair and just manner, taking special account of the needs of the South and of the geographically disadvantaged. Paul Engo who chaired the negotiations put it thus:

> For the industrialized countries, it appeared to be part of the eternal struggle of nations for access to power and wealth. The common good of mankind as a whole was not a priority item on the agenda. They were speaking from positions of technological and economic strength. They seemed to be preoccupied with rivalry among themselves. For the young nations of the Third World, it was a fundamental matter of survival: a struggle for the right to participate in the sharing of benefits, keeping a watchful eye on their worsening economic malaise. The priority item was the collective interest of mankind as a whole from which they held hopes of equity.[37]

The divergence between liberal market and authoritative allocation principles was eventually resolved in a quite novel manner. The compromise entailed an application of the 'I cut, you choose' method of dividing a cake. A prospective seabed miner would have to stake out a claim divided into two equal parts. It would then be up to the international Enterprise (representing the community interest) to choose which part the applicant would mine and which would be reserved for common exploitation by the Enterprise. Such a parallel system combined elements of corporate enterprise and 'common heritage' but required the former to bear the burdens of prospecting and to transfer both technology and some revenue to international bodies. Many other allocative systems were discussed, including joint enterprises between developed and less developed states.[38] The parallel system eventually chosen owed much to American inspiration and represented a major innovation in the international management of commons resources.

Krasner (1985: 234–235) has claimed that 'this is the closest the Third World has come to constructing its ideal regime'. Not only does it provide for central planning, allocation and revenue and technology transfer, but it also enshrines the 'one state one vote' principle in decision-making in the Authority – although there are complicated blocking mechanisms. The Convention also incorporates elements of another NIEO principle, the management of commodity markets. Hence there are provisions to limit seabed mineral production to ensure that there is no adverse effect upon land-based mining industries.

Having said this, towards the end of the negotiations the G77 were prepared to go a long way to meet objections from developed world mining consortia and governments, particularly in granting special status and mining rights to pioneer investors (known to the Americans as 'grandfather rights' and more formally as Pioneer Investor Protection). Before that, negotiation of terms by the US and its allies had not been an act of altruism. As Hollick (1981) and Schmidt (1989) have demonstrated, American policy was very much subject to the 'pulling and hauling' of domestic pluralist and bureaucratic politics. Yet, in the face of these parochial concerns, 'the Nixon, Ford and Carter Administrations appeared committed to the pursuit of the multilateral option' (Schmidt, 1989: 78). This option involved negotiating the compromise of a parallel system and recognized the linkage between the seabed issue and a range of US interests, not least amongst which was the strategic requirement for rights of passage. The reasons for ultimate rejection of the seabed regime by the US, Britain and West Germany were couched in terms of the commercial viability of the system and the precedents set by this type of regime. However, the real reversal of the early 1980s was, most commentators agree, much more profound. It involved the abandonment of the concepts of interdependence that had held sway in the Washington of the 1970s and had served to keep the opposition of mining interests in check. The 'neo-liberal' ideology of free market capitalism and tough defence of Western interests associated with both the Reagan and Thatcher administrations had very little time for the concept of 'common heritage' or for the North–South dialogue in general. Equally, the institutional machinery of the seabed regime was seen as setting a wholly undesirable precedent.[39]

Organization and Procedures

There can be few, if any, examples of an international common property resource regime elaborated in such minute detail. The seabed regime takes up 58 of the 320 Articles in the LoS Convention and is further developed in two extensive annexes. As a very proactive regime its organizational elements are particularly important. The supreme organ is the International Seabed Authority. Under Art. 153 it has the power to organize, carry out and control activities in the 'Area'. It is afforded legal personality and immunities and has the power to borrow funds (Arts 171–183). The Authority comprises an Assembly, Council and an Enterprise, assisted by an international secretariat and in the case of the Council by two specialized Commissions. The Assembly, comprising all state parties to the Convention, operates on a one state one vote basis (with a two-thirds majority requirement for substantive matters). Its executive arm is the Council of 36 state members elected by the Assembly on a proportionate basis reflecting different regions and interests.[40] The Council is charged with establishing specific policies within the general guidelines established by the Convention and the Assembly. Precise functions are specified in great deal in the 26 paragraphs of Art. 162. Voting on the Council is correspondingly complex. Different majorities are required for different issues ranging from a simple majority for procedural items, through three-quarters for matters of implementation and choice between contractors and consensus for the sharing of revenues.[41] As Krasner (1985: 234) comments, 'A major industrialized country would thus have a veto only on those issues decided by consensus, and might not be able to block decisions requiring a three-quarters majority'. The Enterprise is 'the organ of the Authority which shall carry out activities in the Area directly' (Art. 170). This novel international entity, in the sense that it is designed to engage in production itself and to operate autonomously on the basis of 'sound commercial principles', is extensively described in a 'statute' which is Annex IV of the Convention. Its governing body is composed of 15 members elected by the Assembly who will preside over a permanent staff and director general. The organizational picture is completed by a dedicated court, the Seabed Disputes Chamber. The arrangements in the Convention are subject to review 15 years after the first commercial production of seabed minerals. The review would operate under the same voting rules as the LoS Conference itself, requiring consensus. However, if this has not been achieved after a period of 5 years, a three-quarters majority will suffice to revise the Convention (Arts 154–155).

Rules, Information and Enforcement

At the heart of the distributional rules of the seabed regime is the principle of 'parallel' private sector and international Enterprise mining. The detailed mechanism stipulates that each applicant shall submit plans for an area 'sufficiently large and of sufficient estimated commercial value to allow two mining operations' (Annex III(8)). The Authority will then reserve one of these areas for exploitation by the Enterprise or by developing countries. In order to allow the Enterprise to commence operations, it was to receive start up funds, 50% in the form of interest-free loans from members of the Authority and 50% in the form of ordinary interest-bearing loans. In the longer run the Enterprise was to become a self-financing and profit-making concern. More controversially, mining company applicants would not only undertake prospecting for the Enterprise but be under an obligation to transfer relevant technology 'on fair and reasonable commercial terms and conditions' to the Enterprise or to developing countries.[42] This transfer is limited to a period of 10 years after first commercial production from the Area and may only be invoked if alternatives are not commercially available to the Enterprise.

Mining applicants would, under the Convention, also be subject to a form of taxation by the Authority. This, like much else in the Convention, is specified in minute detail and involves two complex systems, one based on royalties related to the market value of production and the other containing both royalty and profit-sharing provisions (Annex III(13)). The Authority would then redistribute these revenues to its members or invest them in the Enterprise. The effort and time devoted to the negotiation of such complex provisions seems, at first sight, puzzling for an industry that had not yet come into existence.

The answer appears to be that developed countries were concerned to avoid any association between access to the seabed and the writing of blank cheques, especially in the light of the voting provisions in the Council, and that the G77 were determined to ensure that the Enterprise would become an operating reality. As if the provisions were not developed and detailed enough, the major function of the Preparatory Commission, set up in Jamaica by the Conference to oversee introduction of the regime, was to elaborate more rules and procedures. It was, as Ogley (1984: 237) comments: 'a monument to the distrust felt among prospective mining states . . . of a Seabed Authority endowed with significant discretion'.

The regime also envisages a degree of commodity market management in the event that seabed mining would prejudice the economic interests of existing land-based mineral producers. The Convention makes the 'promotion of just and stable prices' an objective along with the protection of developing countries from 'adverse effects' caused by activities in the Area.[43] For an interim period of 25 years the Authority will calculate and then impose a seabed production ceiling based upon world output of nickel. Under Art. 151 direct compensatory action to assist developing countries whose export earnings are affected by seabed mining is also envisaged.

The handling of information is a significant feature of any commons regime, but in this case it is fraught with difficulty because of its potential commercial value. Mining contractors are obligated to supply the Authority with detailed information and to transfer all necessary data as to their activities. Such information is divided into two categories: proprietary and non-proprietary.

The first, which relates to the specifics of the contract, is treated as commercially confidential and may not be communicated by the Authority even to the Enterprise (except in relation to reserved areas). The second may relate, for example, to the environmental responsibilities of the Authority under Art. 145 and can be publicized (Annex III(14)). The protection of the seabed environment is also covered in articles assigning responsibility to the Authority.[44] In addition, the Authority, acting through the Council, has rights of investigation and to this end it is envisaged that the Council will direct a staff of inspectors and receive technical advice from the Legal and Technical Commission.[45]

The Authority is invested with extensive enforcement powers involving the right to take 'at any time any measures' to ensure compliance with contracts and the Convention. In this it is assisted by the requirement that all contractors are sponsored by state parties who undertake responsibility for their compliance.[46] Furthermore, the Council, acting as the executive of the Authority, has the right to initiate legal proceedings in cases of non-compliance before a dedicated court, the Seabed Disputes Chamber (Arts 186–191).

Objection to Part XI was the major reason for the refusal of most developed countries, following the US lead, to accede to the whole Convention. A number of them, having an interest in possible mining, have passed national enabling legislation (notably an Act of the US Congress in 1980). Taken together with a 1982 agreement between the USA, France, Germany and the UK, this might constitute an alternative regime. However, as Ogley (1984: 241) observes, the agreement was designed in ways which would allow future operation within the Convention and provides 'not for mutual recognition of mine sites but only for mutual consultations over claims that might overlap'. An embryo or 'interim regime' continued to be supervised by the Preparatory Commission for the International Seabed Authority.

As the number of ratifications of the LoS Convention gradually approached the level required for entry into force, the situation regarding Part XI began to change. From 1990 the UN Secretary General sponsored consultations involving contacts with both developed and less developed countries on 'outstanding issues relating to the deep seabed provisions'. The consequence was an informal draft agreement known as the 'Boat Paper', designed to provide the basis for developed countries to accede to the Convention through the very substantial modification of the seabed regime.[47] If Part XI were operating it would be appropriate to refer to these 'consequent adjustments' as significant regime change. While reaffirming that the Area is still the common heritage of mankind, it is stated that 'political and economic changes, including a growing reliance on market principles show the need to re-evaluate some aspects of the regime'. This was something of an understatement. The provisions contained in a lengthy annex to the draft agreement substantially weaken the independence and viability of the Enterprise. Initial mining operations will be joint ventures and the obligations to transfer technology and fund the Enterprise are removed. In decision-making 'consensus will be the rule'. At the same time Art. 151 is modified to prevent the Authority having a production policy and the system of compensation for LDC mineral producers is deleted. The provisions of Annex III, containing the painstakingly negotiated financial arrangements between contractors and the Authority, are also replaced by a requirement for future negotiation.[48]

These modifications were adopted by the UN General Assembly on 28 July 1994 and entered into force exactly two years later. In November 1996 the Authority commenced operations at its base in Jamaica under its first Secretary General, Satya N. Nandan of Fiji. Most of its initial work was organizational and administrative, establishing a secretariat and taking over the functions of the Preparatory Commission. The Council and Assembly began to function and elected two subsidiary bodies, a Legal and Technical Commission and a Finance Committee.[49] No seabed mining actually occurred during the 1990s but the Authority was engaged in substantive preparations. These involved the drafting of a mining code by the Legal and Technical Commission, the compilation of a secure database on the deep seabed resources of the Area (POLYDAT) and the study of the resources in those parts of the seabed which have been reserved for the Authority. At the same time the Authority began to conduct studies of the possible environmental impact of deep seabed mining of polymetallic nodules. In 1997 the exploratory work plans of seven 'pioneer investors' were approved by the Council and mining will commence once the code has been finalized and contracts issued. It will, however, be on a rather different basis from that originally envisaged for the 'common heritage' regime of the seabed.[50]

Summary

For many years it would have been possible to speak of a single regime for the oceans, or more correctly the high seas. Its principles and norms were simple and permissive, stressing free access and exploitation. The consequence of technological change and industrialization and the attendant stress placed on the ocean commons, was to set up pressure for their enclosure by coastal states. This process was legitimized by UNCLOS III, which also provided a set of overarching norms and principles relating to the remaining ocean commons as well as an elaborate blueprint for a seabed minerals regime. The Law of the Sea covers living resources, but whales along with the fisheries of Antarctica represent the only truly global marine common property resource. The whaling regime provides an example first of the way in which inadequate rules and procedures may lead to a commons tragedy and then of radical regime change spurred on by the environmental movement. Far less glamorous, but of ultimately much greater significance, is the problem of marine pollution. Here, there are what might be best described as partial regimes covering sea-based dumping and pollution, but an absence of effective regulation to cover the most serious problem of land-based pollution, although local and regional regimes may provide the most effective form of governance in this instance and there are signs of a more integrated approach. There has been a notable development and tightening of the existing institutions such as to constitute regime change. The putative seabed regime was in sharp contrast. It contained a pioneering attempt to apply 'common heritage' principles by replacing established doctrines of ownership and providing mechanisms to compensate the disadvantaged for the exploitation of a common resource. It also had decision-making and enforcement mechanisms far removed from the decentralized, nation-state-based forms found in the other maritime regimes. It remained, for more than a decade, little more than an embryo, but as subsequent chapters will demonstrate, a significant one. When it finally entered into force, on 16 November 1994, it was in very significantly weakened form.

Notes

1. The existing Law of the Sea and the incentives for revision are well covered in Barston & Birnie (1980).
2. The provisions in Art. 76 are carefully stated and complex with definitions relating to the outer edge of the continental shelf and depth criteria.
3. *Earth Negotiations Bulletin*, 21 December 1993, 7, 17, p. 2.
4. The 1982 Convention (Art. 65) recognizes that whales are a global common property resource and notes that existing institutional arrangements are not prejudiced by the setting up of EEZs. There are a number of maritime environmental provisions in the Convention. Part XII is devoted to the 'Protection and preservation of the marine environment' and contains in Section 6 references to obligations in respect of existing rules and 'competent international organizations'.
5. Peterson (1993) provides a listing and an analysis of the history and functions of the fisheries commissions.
6. Brown, P. & Vidal, J., 1994, 'Catastrophe Threatens World's Fisheries as Stocks Fall', *Guardian*, 12 March.
7. Agreement for the Implementation of the Provisions of the UN Convention on the Law of the Sea of 10 December 1982 Relating to the Conservation and Management of Straddling Fish Stocks and Highly Migratory Fish Stocks. A/CONF.164/37, 1995.
8. Baleen refers to a fibrous material, which used to be termed 'whalebone', and is formed in plates in the mouth of the whale, functioning as a sieve for planktonic food. Whales equipped and feeding in this way are termed 'baleen' (10 species) as opposed to the 69 species of toothed whales which eat fish and squid.
9. Cited in Birnie (1985) p. 164.
10. Environmental Investigation Agency, 1993, *Whale Wars — The Facts*.
11. The British presented these ideas at the 1993 IWC, pointing out that whale watching constituted the world's fastest growing industry (*Guardian*, 14 May

1993). The figures are from Rossiter, W., 1997, 'Whale Watching World Wide', Cetacean Society International, *Whales Alive!*, Vol. VI, No. 4, October.

12. Cited in Vidal, J., 1993, 'Weeping and Whaling', *Guardian*, 7 May, p. 18.

13. For an in-depth discussion of these normative issues see Stoett (1997: 102–132). The author concludes that although whales are special and 'widely admired for their biological and social characteristics . . . it is difficult – ethically, politically, legally – to deny all human beings everywhere the right to eat them' (Stoett, 1997: 132). At the same time an anti-consumptionist argument based simply upon preservation contains the seeds of its own destruction, because if successful, whale populations will rise.

14. Those states voting against were: Iceland, Japan, Korea (PR), Norway, Peru, USSR and Chile. Source: Birnie (1985: 614).

15. *Guardian*, 10 June 1993.

16. See O'Connell, K., 1997, 'CITES: A Whale of a Meeting', Cetacean Society International, *Whales Alive!*, Vol. VI, No. 3, July. A resolution attacking the IWC moratorium sponsored by Japan and Norway failed by 51 votes against to 27 for, while a Norwegian proposal to downlist minke whales was more narrowly defeated by 57 to 51.

17. 1 BWU = 2 fin whales = 2½ humpbacks = 6 sei whales. Source: Driver (1980: 42–43).

18. The New Management Procedures introduced by the 26th ICW meeting were a response to calls for a moratorium, allowing whaling to continue, but adopting the principle of maximum sustainable yield (MSY). Whale stocks were to be placed in three categories: initial management stocks which could be hunted and reduced to MSY or optimum levels; sustained management stocks which were to be held at or around MSY; and protection stocks which would be completely protected. The determination of the actual levels and which stocks were to go into which category placed a major burden on the Scientific Committee of the ICW.

19. Press reports claimed that a deal had been struck between Brundtland and US Vice President Gore. The NGOs were also badly divided, with Greenpeace supporting the sanctuary plan but other organizations such as the Environmental Investigation Agency refusing to countenance any arrangement which sanctioned the killing of whales (*Guardian*, 27 and 28 May 1994; *Observer*, 24 April and 29 May, 1994).

20. The post-Soviet Russian Government revealed the extent of these breaches to the IWC in 1994 (Brown, P., 1994, 'Soviet Union Illegally Killed Great Whales', *Guardian*, 12 February, p. 12). For a summary of an authoritative Russian report with details of unreported catches in the period 1947–72 of 9000 blue whales, 46 000 humpback and 21 000 sperm whales see Yablokov, A.V., 1997, 'On the Soviet Whaling Falsification, 1947–1972', Cetacean Society International, *Whales Alive!*, Vol. VI, No. 4, October. Yablokov's conclusion is that this 82% deviation from data submitted to the IWC means that 'All previous population models for whales now need to be recalculated'.

21. The relevant legal instruments are the Pelly and Packwood–Magnuson amendments. Japan was subject under these amendments, in 1983, to limitation of its fishing activities within the US 200 mile zone as retribution for its taking of minke whales. The provisions of CITES also apply – whales being classified as endangered species.

22. *Guardian*, 11 May 1993.

23. The 50th meeting of the IWC at Muscat, Oman, May 1998, passed a resolution by consensus to give high priority to the research initiatives of the Standing Working

Group and to make 'environmental concerns' a regular agenda item. See Carson, C., 1998, 'The International Whaling Commission at 50', Cetacean Society International, *Whales Alive!*, Vol. VII, No. 3, July.

24. Cited in Nurmi (1988: 207).
25. Agenda 21, 1992, Ch. 17: B, p. 141.
26. 13th Consultative Meeting of the LDC, London, 1990, quoted in Birnie & Boyle (1992: 319).
27. David Egan, who has specialized in this area, observes that: 'The National Ocean Service in the US is currently the best example of a national agency seeking to collaborate with the regime. It can be seen as the first attempt to link policy and management of land as well as sea-based disposal practices and may, if it succeeds, become analytically significant in terms of the merging of issue areas. It also provides a major contribution to realization of the idea of integrated coastal management and is procedurally significant in linking the agencies of the dumping regime (potentially) to a host of established national agencies with resources and technical expertise that far exceeds that available to the IMO' (personal communication, June 1999).
28. Information on the 14th Consultative Meeting is from Ringius, L., 'Global Environmental Regime Change: Lessons from Ocean Dumping of Radioactive Waste', MS in draft, who argues that this constitutes regime change.
29. An indicator of the scale of activity is provided by secretariat costs: for 1990 for the London Convention these were £452 000 but for MARPOL $3 million. Source: Sand (1992: 157 and 163).
30. Because of the excess tonnage that was created following the fall in demand for oil in the 1980s, large numbers of vessels built before 1978 (with construction not determined by the MARPOL 1973/8 standards) have been laid up or only partially utilized. The accident prevention standards of the 1992 Protocol have made many of these rusting ships non-cost-effective and thus liable to scrapping.
31. Communication from David Egan.
32. Sand (1992: 156).
33. A US National Academy of Sciences 1990 report to IMO cited in Sand (1992: 160). The report indicated that in 1989 568 000 tonnes of oil entered the sea from ships, compared with a figure of 1.47 million tonnes in 1981.
34. A detailed survey of the geological, technological, economic and environmental aspects of deep seabed mining is provided by Ogley (1984, Ch. 2: 4–30). A briefer account with details of the corporations involved is to be found in Sanger (1986: 158–167).
35. Treaty on the Prohibition of the Emplacement of Nuclear Weapons and other Weapons of Mass Destruction on the Sea Bed and Ocean Floor (1971). There are similar commitments relating to other commons areas in the 1959 Antarctic Treaty and the 1967 Outer Space Treaty. According to Schmidt (1989: 25), the superpowers ensured that the issue of seabed arms control which had figured in Pardo's speech was transferred to the Eighteen Nations Disarmament Committee in Geneva.
36. For a discussion of the need for some form of regime and a detailed and authoritative account and analysis of the negotiation of the parallel system see Ogley (1984) Ch. 7.
37. Cited in Sanger (1986: 169).
38. On joint ventures, see Sanger (1986: 169–173).
39. On this see Sebenius (1984: 104) and Schmidt (1989: 214–260).
40. Krasner (1985: 233–234) summarizes the Art. 161 representational arrangements

as follows: 'The Assembly elects the Council, which has thirty-six members chosen from five different classes: four large consumers of metals (including one from Eastern Europe); four major investors in nodule exploitation (including one from Eastern Europe); four land-based producers (including two LDCs); six LDCs with special interests (including landlocked and large populations); and eighteen states from designated geographic regions'.

41. Article 161(7) stipulates that one member shall have one vote. Decisions on procedure will be by majority of those present and voting (8a), decisions of substance on a list of issues, for example reporting and making recommendations, require a two-thirds majority (8b) and for other issues involving implementation, budgeting and rule-making a three-quarters majority (8c). Consensus, defined as the absence of formal objections, is required for measures relating to the protection of producers from adverse economic effects and for the equitable sharing of revenues (8d). In cases of dispute as to which majority applies the higher or highest majority will be deemed to apply (8g).

42. Annex III, Art. 5.

43. Article 150 (f) and (h).

44. Articles 208 and 209 of the Convention contain exhortations on the avoidance of environmental damage but leave it to the Authority to make rules.

45. See Arts 153(a) and 162(2) along with Art. 165 for the Legal and Technical Commission.

46. See Art. 153(4) and (5) and, in relation to sponsoring states, Annex III Art. 4.

47. This section is based upon the working documents: 'Information Note: concerning the Secretary-General's Informal Consultations on outstanding issues relating to the deep seabed mining provisions of the United Nations Convention on the Law of the Sea, New York, 27 and 28 April 1993', 28 April 1993.

48. 'Resolution Adopted by the General Assembly' and 'Agreement Relating to the Implementation of Part XI of the 1982 United Nations Convention on the Law of the Sea', UN General Assembly A/Res/48/263, 17 August 1994.

49. International Seabed Authority, 'Report of the Secretary-General', ISBA/3/A/4, 31 July 1997.

50. International Seabed Authority, 'Report of the Secretary-General', ISBA/4/A/11, 20 July 1998. The seven registered pioneer investors who will be the first to receive contracts and mining licences are: The Government of India; Institut Français de recherche pour l'exploitation de la mer and Association française pour l'étude et la recherche des nodules; Deep Ocean Resources Development Co. Ltd (Japan); Yuzhmorgeologiya (Russian Federation); China Ocean Minerals Research and Development Association; Interoceanmetal Joint Organisation (Bulgaria, Cuba, Czech Republic, Poland, Russian Federation and Slovakia); and the Government of the Republic of Korea.

4

Antarctica

Antarctica forms some 10% of the earth's land surface. Its remote, frozen and inhospitable character explains why it was the last continent to be explored. Although commercial sealing occurred off its shores during the nineteenth century, it was only in the early years of the twentieth that significant exploration began. Even today the various scientific stations that have been established hardly constitute permanent human habitation as normally understood. Antarctica remains a commons because the seven national claims to sovereignty that have been made, by Argentina, Australia, Chile, France, New Zealand, Norway and the United Kingdom, have not become a territorial reality. Whether based on discovery, geographical propinquity or geological continuity they fail the decisive test of continuous occupation. As Joyner (1992: 66) explains: 'Occupation as the basis for acquiring sovereignty in Antarctica must remain the essential condition for effecting that sovereignty there'. The foundation of the continent's legal status and of the Antarctic regime is Art. 4 of the 1959 Antarctic Treaty which puts the various territorial claims into abeyance. Antarctica thus effectively remains a commons (at least for the duration of the Treaty). This status is extended to the surrounding Southern or Circumpolar Ocean. Clearly, in relation to the Law of the Sea, Antarctic waters cannot be nationally appropriated in a territorial sea or an exclusive economic zone (EEZ) if there is no territorial sovereignty to provide the basis for such claims.[1] In fact, rather than falling under general high seas provisions, part of the Southern Ocean is explicitly placed under the Antarctic Treaty System (ATS) for the purposes of conserving marine living resources.[2] For the purposes of the Convention on the Conservation of Antarctic Marine Living Resources (but not for the Antarctic Treaty itself) the ocean boundary of Antarctica is physically defined as the Antarctic Convergence (Figure 4.1). This is marked by the meeting of the northward-flowing waters from the Antarctic land mass and the subtropical waters of the South Atlantic. In the Convergence the cold waters fall beneath the warmer and the associated turbulence brings nutrient-rich water

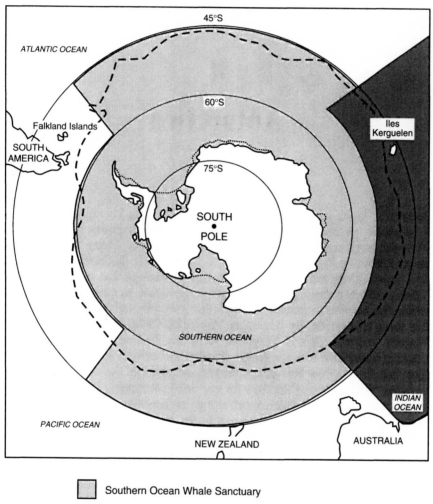

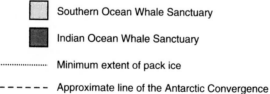

Southern Ocean Whale Sanctuary

Indian Ocean Whale Sanctuary

Minimum extent of pack ice

Approximate line of the Antarctic Convergence

Figure 4.1   *Antarctica: Treaty limits and the Whale Sanctuaries*

to the surface. During the Antarctic summer this provides the basis for abundant growth of phytoplankton which turns the Southern Ocean into one of the richest of all maritime feeding grounds. Biomass is actually greater than that existing in tropical waters, although fewer species are to be found. These include seals, whales, fin fish such as the Antarctic cod, and, at the base of the food chain, the shrimp-like krill. This abundance is in stark contrast to the stunted range of flora and fauna, mosses, lichens and insects, to be found on the ice-covered continent itself. Maritime species thus continue to constitute the main commercially exploitable resource of Antarctica.

Their exploitation has provided some classic instances of the 'tragedy of the commons'. From 1790 and onwards throughout the nineteenth century there was intensive commercial sealing activity in the Southern Ocean. In the search for skins and oil literally millions of Antarctic fur and elephant seals were killed and by the 1920s the industry had essentially destroyed itself by bringing seal populations to the brink of extinction. Some of the hunters, like Captain James Weddell, 'recognised that they were destroying a resource which with more prudence might have yielded a steady return indefinitely [but] it was in nobody's interest to exercise restraint' (Holdgate, 1987: 129). During the first half of the twentieth century the Southern Ocean was the location for one of the whaling 'tragedies' mentioned above – the predictable result of an open-access fishery (Gulland, 1987: 117). The majority of baleen whales (including the blue, minke, humpback, sei and southern right whales) depend upon a seasonal migration to the Southern Ocean where they 'filter feed' off the krill and plankton. This demonstrates the importance of the International Whaling Commission's Southern Ocean sanctuary, agreed in 1994 (see Figure 4.1).

Commercial fishing in the Southern Ocean is a relatively recent phenomenon – dating from the 1960s. It was given a major impetus by the developments in the Law of the Sea towards enclosure through EEZs. They effectively debarred the Soviet and other distant-water fleets from many of their traditional fishing grounds, causing them to look southwards. The primary species taken are from the Antarctic cod (or Nototheniidae) family. There is also an emergent krill fishery for, although the main species, *Euphausia superba*, is only 2–4 cm in length, the biomass of krill is so large (greater than any other species on earth) that it is seen as a major potential food source for a hungry world if and when other fisheries fail. (However, direct human consumption depends on finding a commercial method of making it palatable.) It was an awareness of these developments, in the light of the 'tragedies' experienced with seals and whales, which prompted the formulation of the Convention on the Conservation of Antarctic Marine Living Resources (CCAMLR, 1982). There is special concern for the generally acknowledged fragility of Antarctic marine populations and the fundamental importance of the humble krill which supports an unusually short food chain.

The other resources of Antarctica are more potential than actual. There are mineral and hydrocarbon deposits, although their extent is subject to dispute and the sheer technical and logistical difficulties of extraction mean that they would not represent a commercial proposition without very major increases in world market prices. The situation is, perhaps, a rather more extreme case of that encountered with seabed minerals. This did not, however, prevent the expenditure of a great deal of time and diplomatic energy, during the 1980s, in the negotiation of a minerals convention (CRAMRA). It was ultimately to be stillborn because of the refusal in 1988–89 of two Antarctic Treaty Consultative Parties, Australia and France, to ratify.

Around three-quarters of the world's total supply of fresh water is trapped in the polar ice caps. At some future date this may represent an exploitable resource and the legal status of icebergs as a common property resource is a topic for speculation which is not entirely fanciful.[3] The other emergent commercial activity in Antarctica is tourism. Although at the moment the industry exists on a small scale involving approximately 10 000 visitors per annum (Davis, 1998: 45), it has already been a cause of some concern relating to the environmental damage which may be occasioned, particularly at sites of special scientific interest and popular tourist destinations such as Port Lockroy on the Antarctic Peninsula. The tourist industry, in the shape of the International Association of Antarctic Tour Operators and the Pacific Asia Travel Association, is now a regular presence at Antarctic Treaty Consultative Meetings (Beck & Dodds, 1998: 40).

From the beginning of Antarctic exploration, scientific interest has accompanied the search for resources. Both are inextricably entwined. Given the geophysical significance of the poles, Antarctica provides a unique setting for scientific research and a 'simple' environment largely unaffected by human activities. The importance of international collaborative research in polar science was recognized by the International Council of Scientific Unions (ICSU) in setting up the International Geophysical Year (IGY) of 1957/58. As well as stimulating major advances in fields such as geomagnetism, meteorology and glaciology, it also provided the principal stimulus for the negotiation of the Antarctic Treaty of 1959, which continues to provide the foundation of the Antarctic regime. The occupation of the continent by human settlements has had an expressly scientific purpose and, as will be seen, the active conduct of such research provides the necessary basis for full national participation in the regime.

Recent research has emphasized the interconnection between Antarctica and global environmental change. There is increasing understanding of the significant role played by Antarctica in the global climate system. The high albedo of the ice cap gives the continent 'a world wide influence due to the fact that it is the strongest cooling centre in the global system' (Laws, 1987: 28). Temperature gradients between high and low latitudes provide the thermal drive for global circulation. There is also the question of the stability of parts of the ice sheet under conditions of global warming. As well as being an important influence on climate in its own right, Antarctica also provides an essential laboratory for the study of climate change. The drilling of ice cores can provide a record of atmospheric changes over a period of 100 000 years and the remoteness of Antarctica also means that it provides a baseline for studies of global pollution levels. Antarctica provided the site for the definitive investigation of the depletion of the ozone layer and notably the Farman study of 1985 (funded by the British Antarctic Survey). This was in part occasioned by the very special conditions of the stratosphere over the poles which accentuated ozone loss and also with the purity of the Antarctic atmosphere that facilitated observation.

The scientific and resource dimensions come together in the recognition that commercial and indeed scientific activities may damage the very special and uniquely vulnerable environment of Antarctica. It is not only that such degradation would reduce the scientific utility of the continent as an unspoiled laboratory but also that, in a deeper sense, Antarctica represents the last great wilderness on earth. Thus non-governmental organizations and even some governments have pressed the case for realization of 'wilderness values' by the transformation of Antarctica into a 'world park'. This was linked in practical and political terms with the successful campaign for a Southern Ocean Whale Sanctuary – agreed by the IWC at its 1994 meeting.

The Antarctic Treaty System – A Single Regime

The arrangements governing the activities described above are normally termed the Antarctic Treaty System (Vidas, 1996). As far as legal instruments are concerned the system comprises the 1959 Treaty itself, a large number of agreed measures adopted by the parties to the treaty and the 1991 Madrid Protocol relating to environmental protection. Maritime resource issues are regulated in association with the Treaty by the 1972 Convention on the Conservation of Antarctic Seals (CCAS) and the 1982 Convention on the Conservation of Antarctic Marine Living Resources (CCAMLR). An additional Convention on the Regulation of Antarctic Mineral Resource Activities (CRAMRA) was painstakingly negotiated during the 1980s leading to the production of an agreed text in 1988. The CRAMRA was vigorously opposed by environmental NGOs who since the 1970s had been raising public awareness of the potential desecration of this last great wilderness. They were joined by two of the ATS states, Australia and France. Having both agreed the Convention they proceeded in 1989 to breach the usual inter-governmental consensus in Antarctic politics by a refusal to accede. This dramatic reversal of position led to a wholesale re-evaluation of both the mining question and the role of the ATS in environmental conservation. In consequence the 1991 Madrid Protocol (also known as the PREP – Protocol on Environmental Protection), comprising not only formal rules for environmental impact assessment but also a permanent ban on mining, was speedily negotiated. The Protocol, which entered into force in January 1998, underlines a significant shift in the character of regime from an original political/territorial accommodation towards comprehensive environmental protection.[4]

Commentators frequently refer to the whole Antarctic Treaty System as a regime while at the same time referring to specific maritime (CCAMLR) or mining (CRAMRA) regimes.[5] It has already been noted that despite the physical unity of space or the oceans, issue areas and hence regimes do not often reflect such holistic conceptions. Rather they are fragmented in ways which accord with more specific activities and concerns. The attempt to consolidate the various maritime regimes within a single all-encompassing Law of the Sea, established some important norms and principles and significant political linkages, but as argued in the previous chapter, hardly constitutes a single integrated regime. However, this is emphatically not the case with Antarctica. The governments that have been involved since 1959 in the regulation of the Antarctic, principally the Antarctic Treaty Consultative Parties (ATCPs), have tended to treat the various resource scientific and other Antarctic matters as a single issue area. This has been reflected in a continuity of personnel across the various negotiations and above all in the legal architecture which has been erected on the foundation of the 1959 Treaty. Such an approach has clear advantages in that questions of principle do not have to be re-negotiated every time a new set of issues are addressed. Above all, it allows for a comprehensive approach where regulatory activity can reflect actual physical circumstances and problems such as those that exist in the relationship between resource extraction and the protection of the environment.

Principles and Norms

Some have doubted whether the Antarctic Treaty System constitutes a regime at all. In Krasner's view we are confronted with a 'logical shambles'. Fundamental differences of principle have not been resolved while a set of 'rules and decision-making procedures, reflecting weak norms associated with scientific exploration, disarmament and environmental protection' have been put in place in order to avoid 'mutually undesirable outcomes' (Krasner, 1985: 251). On the other hand, Falk (1991: 399) claims that, 'The governance of Antarctica, ingeniously combining the virtues of international cooperation with the reality of state sovereignty, is the closest thing to a "world order miracle" that the world has known'. The regime has evolved incrementally and to a large degree consensually, in ways that could not have been foreseen by its founders and which have entailed some significant shifts at the level of norms and principles, alongside very extensive development of specific rules.

The original treaty, concluded in the midst of the Cold War, was seen as a means of continuing the Antarctic research activities associated with the 1957/58 IGY, while avoiding territorial and security conflicts. It contained two norms of behaviour. The first under Arts I and V stipulated that the continent was to be used for peaceful purposes alone and prohibited nuclear explosions and the disposal of radioactive wastes. The second, under Arts II and III urged the promotion and facilitation of collaborative international scientific research, making the IGY effort a permanent fixture. The main device for achieving this was the principle (or perhaps non-principle) of avoiding the question of ownership of Antarctica by putting the territorial demands of the seven claimant states into abeyance. Under the celebrated Art. IV of the 1959 Treaty, nothing in the text was to prejudice existing claims or constitute a basis for asserting them. Neither were any new claims or enlargements of existing claims to be countenanced during the lifetime of the Treaty. This not only avoided immediate conflicts about overlapping claims but also dealt with the objections of the United States and Soviet Union, both of which rejected all existing claims to the continent while reserving the right to make their own.[6] Article IV has inspired extensive legal gymnastics and essentially enshrines an agreement to disagree or rather to set aside disagreements for the time being. On occasion Treaty Parties have contented themselves with an understanding that actions could be subject to different interpretations in relation to sovereignty – the so-called doctrine of 'bifocalism'. The sovereignty issue was hardly resolved by Art. IV; instead it has continued to lurk never far from the surface of negotiations throughout the life of the Antarctic regime. As Beck (1991: 240) has written, at 'every stage of the regime's development an internal accommodation was required between a range of interests, most notably between those of claimants and non-claimants respecting jurisdiction and the distribution of resource benefits'.

What then is the status of Antarctica? Ambiguity extends beyond the continent itself to the surrounding waters and seabed because the existing law of the sea on EEZs and continental shelves assumes an adjacent territorial sovereign. One obvious potential principle, explicitly and emphatically rejected by the ATCPs, is that Antarctica should constitute part of the 'common heritage of mankind'. During UNCLOS III a Sri Lankan delegate proposed, in 1975, that Antarctica should be considered alongside the deep seabed. This initiative was quickly stifled by diplomatic pressure from leading ATCPs and set aside until the conclusion of UNCLOS III (Suter, 1991: 72–73). In 1982 the Malaysian Government placed the governance of Antarctica on the agenda of the UN General Assembly and by implication the question of future 'common heritage' status. Consideration of the status of Antarctica became an annual Assembly event in the 1980s but strenuous efforts were made by the ATCPs to ensure that no substantive changes occurred and that the Antarctic Treaty System remained separate and disconnected from the central activities of the UN (but not from specific programmes and specialized agencies). This extended to a boycott of General Assembly votes on Antarctica. However, in 1994 a compromise was reached whereby the ATCPs would report to the General Assembly on a triennial basis. A significant part of this compromise was an invitation to the UNEP Executive Director to attend ATCMs, thereby establishing a link which helps to 'give the appearance of public accountability and greater transparency by the ATCPs to the broader international community' (Beck, 1998: 43).

The influence of the ATCPs was also to be seen in the UNCED project, initiated by the General Assembly in the late 1980s and coming to fruition in the Rio 'Earth Summit' of 1992. Antarctica is of primary importance for the circulation mechanisms that determine global climate and a key site for climate change research as well. It is also a fragile ecosystem of world importance and very significant maritime resources and habitats are contained within the Treaty area. *Agenda 21*, the vast document prepared over two years prior to the Conference, covers in its 40 chapters the whole range of local and global environmental concerns, but with hardly a reference to Antarctica.

For international lawyers there are good reasons why Antarctica, unlike the deep seabed or outer space, may not be accorded 'common heritage' status. Their inaccessibility meant that they were not subject to claims to sovereignty before their designation as *res communis*.

> In marked contrast, Antarctica, prior to the intrusion of the notion of a common heritage of mankind, has been viewed as *terra nullius* by the international community and treated as such by states asserting territorial claims. Thus, 90% of Antarctica has been the subject of serious territorial claims for approximately 60 years. The prevailing situation between states at the time when the notion of a common heritage illumined international consciousness, is thus significantly different from the deep seabed and outer space. (Triggs, 1987: 103)

The political realities are perhaps more significant. On the one hand a large number of less developed countries (LDCs) who were not members of the Treaty system could regard it as a legacy of colonialism and had an interest in trying to bring the Antarctic under some form of UN-sponsored control that would, as in the case of geostationary orbit or the seabed, incorporate 'common heritage' principles. The parallels with the deep seabed seemed highly salient once the existence of mineral resources began to be discussed and the ATCPs themselves began, in the early 1980s, to negotiate rules for mining. However, the South could not be at one on this issue and the ATCPs comprised an unusually powerful bloc, uniting both East and West. Argentina and Chile were staunch original participants and defenders of the existing system – and were joined by India, China and a number of other prominent LDCs during the 1980s. The ATCP countries, significantly strengthened in this way, were able to make the claim that in practical terms the existing system had functioned effectively and should, on pragmatic grounds, be left alone.

The 'common heritage' idea also implies universality of membership and rights of use. As opponents frequently pointed out, a fundamental principle of the Antarctic regime was its exclusivity. Full participation in Antarctic decision-making, in the Meetings of the ATCPs, is restricted to the 12 original members (including all the claimants) and any selected additional members (currently 15). The test for admission is the so-called 'activity criterion' of Art. IX of the 1959 Treaty. States are entitled to participate during such time as they demonstrate 'interest in Antarctica by conducting substantial research activity there'. There are also 16 Non-Consultative Parties, which have only recently been granted observer status as a concession related to the UN debates, making a grand total of 43 signatories to the Antarctic Treaty.

The 'self-conferred' rights of ATCPs are not only exclusive, but also in the view of Malaysia's Permanent United Nations Representative, both total and unaccountable.

> under the Treaty, the ATCPs – and they alone – have the rights to make decisions ('exclusive'), and that the ATS assert rights to regulate all activities in Antarctica ('total'), and that decisions within the ATS are not subject to review ('unaccountable'). (Zain-Azraai, 1987: 212)

While these rights are justified by the ATCPs on grounds of the special expertise of members as participants in Antarctic scientific activity, they cannot, according to Zain-Azraai, accord with the legitimate interests of the rest of the international system in an area of such evident global significance. Instead they fly in the face of the democratization of international relations that has been the prevailing trend elsewhere (Zain-Azraai, 1987: 215).

The pragmatic response is that the system continues to 'work' and as Krasner (1985) and Beeby (1991) have argued, continues to work precisely because decision-making is in the hands of a restricted group of states of manageable proportions which have a direct stake in and control over the issues. Although it is not formally stated, members of the Antarctic club have a 'deeply held conviction of the essential unmanageability' of arrangements involving the wider international community, as opposed to their own 'demonstrated expertise' (Haron, 1991: 301–302). The New Zealander who chaired the CRAMRA negotiations speaks of experience demonstrating that once the number of participants 'exceeds somewhere between 10 and 15, it is much less easy to generate a dynamic that will leads to a consensus acceptable to all (Beeby, 1991: 14). However, this advantage may already be disappearing. The response of the ATCPs to attacks on the exclusivity of the regime during the 1980s was to expand membership, improve access to information and to stress the openness of the ATS to interested new participants. These were to include both the NGOs and UNEP.

Joyner (1992: 271) arrives at the conclusion, common amongst close students of the regime, that as long as the present exclusive principles of the ATS protect the Antarctic 'for the interest of all mankind' one is drawn to the 'compelling conclusion' that replacement with an 'untested philosophical notion is neither politically feasible or legally desirable'. The actual position is that the governance of the Antarctic may be described as a *de facto* if not a *de jure* condominion.

Although the attempt to give the Antarctic 'common heritage' status clearly failed in the 1980s, an alternative and rather different concept of a 'world park' began to gain currency. Whereas 'common heritage', as understood during the seabed negotiations, was centrally concerned with equitable shares in the profits derived from the extraction of common property resources, a 'world park' stresses the primacy of conservation. The idea was first mooted in vague form in the early 1970s but was taken up, briefly by New Zealand in 1975 and then by the NGOs organized in ASOC (The Antarctic and Southern Ocean Coalition) and the IUCN (International Union for the Conservation of Nature). The call for 'world park' status was stimulated by increasing tourism, development of Antarctic bases to meet the 'activity criterion' and above all by the mineral surveys from 1973 to 1974 which led to the initiation of the CRAMRA negotiations. The concept was derided by some ATCMs as lacking in content. However, as developed by Greenpeace in particular, which established a World Park Base in Antarctica from 1987 to 1992, its principles, if not its full institutional form, became fairly clear. At its heart was the notion of the primacy of 'wilderness values', that the Antarctic species and their environment have an intrinsic aesthetic and scientific importance that must override all other considerations. It follows that Antarctic wildlife and their habitats must be completely protected.[7] In the debates over CRAMRA, 'world park' principles involving the absolute primacy of complete environmental protection were vigorously promoted by the NGOs as the green alternative to profit-driven mineral extraction. There were both philosophical and practical political connections between this struggle and that which was waged in the IWC over the desirability of sustainable whaling. Arguably, after the collapse of CRAMRA and the negotiation of the Madrid Protocol the world park concept had served its purpose in shifting the agenda towards conservation. Greenpeace closed its World Park Base and attention focused on developing and implementing the PREP (Herr, 1996: 110).

The text of the 1959 Treaty was largely silent upon environmental and resource questions, but these were to dominate subsequent developments. Gradually a substantial body of specific environmental recommendations was developed by the ATCPs, reflecting a largely implicit understanding about the special importance and fragility of Antarctic habitats and norms of environmental good behaviour. In the same period a great deal of effort was expended on developing a regulatory framework for resource extraction and coping with the often conflictual relationship to conservation principles. The 1982 CCAMLR proceeds on the basis of the principle of 'rational use' of marine resources which would maintain 'the ecological relationship between harvested, dependent and related populations' (Art. II CCAMLR). A further attempt to allow resource extraction while protecting the Antarctic environment was painstakingly negotiated during the 1980s in the form of a minerals convention (CRAMRA). The principle of conservation through rational exploitation was not uncontested. NGOs and eventually some governments (Australia and France refused to accede to the CRAMRA) argued that the two concepts were essentially incompatible in the Antarctic context. The incorporation of a complete mining ban (over the 50 year life of the agreement) in the subsequent Madrid Protocol suggests that the principle of conservation through the avoidance of resource extraction has made some headway, but is a long way from gaining acceptance in the area of marine resources. During the Madrid Protocol negotiations some governments were unwilling to forego the option of mining in perpetuity and there was disagreement over exactly how easy it would be to override the ban if changes in the feasibility and economics of resource extraction were to provide incentives.[8]

The Madrid Protocol finally entered into force in January 1998. Its provisions are significant for the future status of Antarctica and for the norms and principles of the regime. Article 2 of the Protocol designates Antarctica as a 'natural reserve, devoted to peace and science' and Art. 3(1) states that:

> The protection of the Antarctic environment and dependent and associated ecosystems and the intrinsic value of Antarctica, including its wilderness and aesthetic values and its value as an area for the conduct of scientific research, in particular research essential to understanding the global environment, shall be fundamental considerations in the planning and conduct of all activities in the Antarctic Treaty area.

The character of this commitment is in marked contrast to the very limited objectives of the 1959 Treaty. The extent of the change is reflected in the way in which norms of environmental responsibility, strict conservation and the need for precautionary action have developed and find their fullest expression in the Protocol.

Organization and Procedures

The Antarctic Treaty established no permanent organizations – indeed their creation would have been regarded by some territorial claimants as a form of internationalization prejudicial to their claims. Instead the 'collective choice' function is exercised by Meetings of the Consultative Parties (ATCMs). Annual meetings are held by rotation in the various member countries. For a long period this was a closed and bureaucratic world populated by specialists from the foreign offices of the Parties (Elliott, 1994: 57). Meetings were held in secret, but from 1990 this rule was relaxed and NGOs were allowed to attend as observers. Since 1978 some 200 NGOs from 40 countries have been united in the Antarctic and Southern Ocean Coalition (ASOC). In 1990 ASOC itself was given invited expert status at ATCMs while a small number of national delegations had already incorporated individual NGO members. It is widely agreed that the role of NGOs at ATCM meetings and outside was critical to the defeat of CRAMRA and the negotiation of the Madrid Protocol (Elliott, 1994: 194; Wapner, 1996: 136–137). Essentially they succeeded through a strategy which not only bombarded the Parties with information but which linked the somewhat closed world of Antarctic decision-making to wider public concerns about environmental degradation.

Uniquely amongst the treaty and convention systems covering the global commons there is still no permanent secretariat and coordinating functions are carried out by the host Government of that year's ATCM.[9] Although there is a minimum of formal organization the years since 1959 have seen the *ad hoc* evolution of a range of specialist committees and commissions that service the ATCMs, such as COMNAP (the Committee of Managers of National Antarctic Programmes). The Madrid Protocol sets up a Committee for Environmental Protection (Arts 11 and 12), but without independent powers or a dedicated secretariat. Particular roles, such as that of Australia in acting as diplomatic coordinator in relation to the UN, have also been defined. Furthermore it has been found necessary to hold Special Consultative Meetings to handle business between regular ATCMs.

The ATCMs make 'recommendations' (Art. IX), and modifications or amendments to the Treaty require not only consensus, but unanimous ratification as well. Such procedures tend towards inaction or agreements that reflect the 'lowest common denominator' and were evidently suited to a regime initially committed to maintaining the political and security status quo. The Madrid Protocol modifies the rules somewhat in the environmental area by allowing that specific amendments agreed by ATCMs will become binding within a year unless one or more parties object.[10]

The functions that were envisaged for CCAMLR required that there be a higher level of formal organization than that pertaining to the ATCMs. *Ad hoc* arrangements simply would not suffice for the detailed business of fisheries management and conservation. Thus, CCAMLR has a permanent Commission of representatives of the Parties, assisted by its own Scientific Committee, both based at Hobart, Tasmania. The Commission is however bound by the same rules of procedure as the Meeting of the Parties. The ill-fated minerals Convention would also have had its own permanent organization and in addition 'a mix of procedures for making decisions, some consensus, some majority and some qualified majority' (Beeby, 1991: 12). In the future functional requirements associated with the new comprehensive environmental role that is envisaged in the Madrid Protocol may force a higher level of formal organization. But for the moment, as Messer & Breth (1991: 388) note, there are still parties who argue from a traditional standpoint where:

> ATS is perceived as a forum which administers Antarctica through gentlemen's agreements. They fear that the establishment of an infrastructure would distort this co-ordination process or take it out of their hands. These misgivings must be understood against the background of the phenomenon of claims to sovereignty in Antarctica. From this point of view, for some countries pursuing such claims, agreeing to set up a secretariat could seem to be a first step towards weakening this claim or giving up part of their sovereignty.

Rules

What the regime may lack in terms of formalized norms and principles, it makes up for in the range and volume of its rules. The specific prohibitions in the original 1959 Treaty cover all military activity (Art. 1) and nuclear explosions and the disposal of radioactive waste (Art. 2). There were no environmental obligations contained in the 1959 text, but the development of the ATS through the Consultative Meetings of the Parties was to see the creation of over 200 Recommendations covering almost the whole range of human activity on and around the Antarctic continent. Attempts at environmental regulation commenced with the adoption in 1964 of the Agreed Measures for the Conservation of Antarctic Flora and Fauna, prohibiting 'without a permit the killing, capturing or molesting of any mammal or bird native to Antarctica by any citizen of any of the Treaty Contracting Parties' (Triggs, 1987: 133). Measures were further developed on codes of conduct for scientific exploration, the running of Antarctic bases and tourism – all designed to minimize human impacts upon a near pristine ecosystem. However laudable the intent, such measures remained recommendations and there is substantial evidence, garnered by NGO observers, that they have been widely ignored, particularly in terms of the pollution associated with permanent national scientific stations on the continent.[11]

Antarctic resources were first protected under a free-standing Convention for the Conservation of Antarctic Seals, concluded in 1972. Stimulated by exploratory sealing expeditions in the mid-1960s, the Convention defined a hunting season and zones combined with very conservative catch limits and a permit system. Whether or not as a result of the Convention, the seal population has remained large and there has been little or no commercial exploitation, except for a Soviet expedition in 1987 which took over 5000 crabeater seals on the pretext that they were for scientific research or for display as museum specimens (Suter, 1991: 32–33).

The Seals Convention was followed by a much more ambitious and ecologically integrated attempt at resource regulation and conservation, the CCAMLR of 1982. The issues to be addressed were the classic ones involved in the regulation of any open-access fishery, but with the proviso that it would not be enough to institute rules for close seasons, mesh sizes and total allowable catches (TACs) for individual species. Instead, because of the very short food chain entirely based on the krill, an integrated approach was required if conservation of marine resources was to be achieved. Thus, under the Convention, parties agree to conduct harvesting and associated activities so as to prevent the decline in stocks but also to maintain 'ecological relationships between harvested, dependent and related populations' and the prevention or minimization of the risk of 'changes in the marine ecosystem which are not potentially reversible over two or three decades' (Art. II). The Convention contains no specific conservation measures, only a commitment by the parties to implement such measures as shall be adopted by the Commission (Art. IX(b)) and to provide data on their harvesting activities (Art. XX). Partly because of the problems of achieving reliable data and consensus amongst members involved in the Antarctic fishery, it took until 1987 to draft significant rules controlling the Antarctic cod fishery and establishing TACs for other species. Up until 1991, 43 conservation measures and eight non-binding resolutions had been established by the Commission (Joyner, 1992: 238). Regarding overfishing, the depletion of the Antarctic cod stocks presented the most immediate problem and the major krill fishery predicted for the 1980s did not materialize. In terms of the ecosystemic objectives of the Convention this was probably fortunate, it having taken until 1991 to establish a precautionary catch limit of 1.5 million tonnes for krill (Joyner, 1992: 238–248). Despite the novelty of its ecosystemic approach, the CCAMLR remains a normal type of fisheries regime employing well established and quite permissive rules. It shares with the whaling regime the characteristic that most of its members are not active 'harvesters' of the resource (Stokke, 1996: 127). There are no national catch quotas and it is arguable that the most effective form of stock conservation in the Southern Ocean has been provided by nationally imposed controls in the waters around South Georgia and the French Kerguelen Islands (Stokke, 1996: 149). There are no provisions in the CCAMLR to deal with non-parties (although they are now invited to attend meetings) or to establish any compensatory arrangements for those unable to profit from the fishery.

In one way it was surprising that the attempt to regulate Antarctic minerals extraction through the CRAMRA was made at all. It represented a rather extreme application of what a participant, Sir Arthur Watts (1992: 224) has called the 'prophylactic provision for resource management' – already identified as a principle of the Antarctic regime as a whole. Although there had been interest in natural gas discoveries off the Ross Ice Shelf in 1973, the overwhelming conclusion of most authorities was that uncertainty over the existence of mineral deposits allied to the extreme difficulties of climate, terrain and transport, effectively ruled out an Antarctic minerals industry for the foreseeable future. Even possible offshore oil and gas extraction was regarded as a prospect unlikely to come to fruition before the latter part of the next century.[12]

Notwithstanding, the question of minerals regulation was placed on the agenda by the 1977 ATCP meeting and led to the negotiation of a framework of rules in Special Consultative Meetings from 1982 until conclusion of a Convention (CRAMRA) at Wellington in 1988. This was assisted by the fact that the parties had, in 1977, come to an informal understanding that prior to the creation of a regime there would be a moratorium on prospecting. The very mention of mining obviously threatened to raise the question of sovereignty and territorial claims. Once again the underlying culture of the ATS was evident in a concern by ATCPs to avoid political conflict without yielding ultimate claims to territorial sovereignty. A central achievement of the negotiations was to work out an agreed solution based on an organizational mechanism specifically involving claimant states on the regulatory committees that would determine particular applications to explore and mine (Arts 9 and 29). The Convention prohibited any Antarctic mineral resource activities outside its rules and procedures (Art. 3). It further attempted to erect strict criteria for judging the acceptability of proposed mining activities in terms of environmental impact assessment and implications for other, and particularly scientific, activities on and around the continent. A framework of rules for the processes of exploration and exploitation was also elaborated in some detail, but it was acknowledged that this did not amount to a mining code and would require further development. The Convention was genuinely guided by the need to protect the Antarctic environment and associated ecosystems, but was severely criticized for the imprecision of its terms (for example the meaning of 'significant adverse effects' or 'substantial risk' in Art. 4 which covers principles concerning judgements on mineral activities). Above all, its very existence implied that mining was a legitimate activity albeit hedged around with restrictions and the need for consensus amongst the parties in designating an area for exploration. It might even be construed as an incentive to mine on account of the promise of possible security of title to a claim. For such reasons, an Australian minister spoke for many in the NGOs and elsewhere in urging rejection: 'we are not convinced that the Minerals Convention would be able to keep intact all the values that it seeks to protect. If we all agree that the environment and other values are more important than mineral resources let us say so now' (Brown, 1991: 115).

In the primacy afforded to environmental protection the CRAMRA differs from that other attempt at regulation of the exploitation of minerals existing beyond national jurisdiction, the seabed regime. The latter was concerned with an activity that appeared at the time to be a much more immediate and real possibility and the elaboration and detail of the financial and other rules reflected this. The central problem was in devising rules for the fair remuneration of the disadvantaged members of the international system for the exploitation of a common heritage resource. The CRAMRA, by contrast, merely spoke in aspirational terms about the 'interests of the international community as a whole' (Art. 2.3(g)). There is a small reference to the distribution of revenues by the Commission, where amounts surplus to its operational expenses will be devoted to scientific research with particular assistance being given to developing country Parties (Art. 35.7(a)).

Significantly, a decade after the seabed debates, the Non-Aligned Movement sponsored a draft resolution in the UN General Assembly which, in condemning the CRAMRA, referred not to the equitable division of potential common heritage mineral resources, but to the need for a total ban on mining and the establishment of a World Park to ensure the protection and conservation of the Antarctic environment 'for the benefit of all mankind' (Barnes, 1991: 215).

The most recent exercise in rule-making is represented by the Madrid Protocol to the original Antarctic Treaty. It marks a 'qualitative change in the approach to environmental issues in the Antarctic and replaces the [previous] *ad hoc* and unwieldy network of measures' (Elliott, 1994: 196). Its Art. 7 treats the mining issue, that had dominated the previous decade, by prohibiting any activity relating to mineral resources other than scientific research within the 50 year lifetime of the agreement. Neither is a World Park established; instead there is a commitment to 'comprehensive protection of the the Antarctic and its dependent and associated ecosystems' (Art. 2). In effect the PREP is equivalent to the framework/protocol model currently utilized in the international protection of the atmosphere and elsewhere (Vicuna, 1996: 190). The rules are contained within five annexes, which consolidate the 'soft law' environmental Recommendations that have been produced since 1964 and give them a legally binding character.[13] It is envisaged that the development of the rules will be 'science driven' and that new annexes (for example on liability) will be added as required and an annex on tourism has been discussed. In general all activities in Antarctica must be planned and conducted so as to limit adverse environmental impacts and to avoid a range of specific forms of degradation (Art. 3). Furthermore, environmental impact assessment for new activities is made mandatory (Art. 8) as is contingency planning in regard to possible environmentally degrading 'emergencies' (Art 15).

Monitoring and Enforcement

The Antarctic regime has lacked any central agency to monitor and enforce its rules. Until recently, it has also avoided binding regulations in favour of Recommendations. The 1959 Treaty set up a system of 'on site inspection' by national observers who would have complete freedom of access to the Antarctic stations of other powers. This was buttressed by an explicit reference to the right of aerial surveillance and a requirement to inform other parties of expeditions, new stations or the employment of military personnel (Art. VII). The inspection and compliance system for the environmental Recommendations made under the Treaty have not been an outstanding success. Formal enforcement is exclusively in the hands of national parties, who are expected to raise questions of non-compliance by others at Consultative Parties Meetings and to discipline their own nationals. There has been an understandable desire amongst the Treaty Parties to avoid giving offence to their counterparts. Reports are described as 'vacuous' and difficulties are 'avoided or glossed over' (Barnes, 1991: 217). A case in point was provided by the 1989 wreck of the Argentinean vessel *Bahia Paraiso* in the Bismarck Strait. Governments studiously failed to address the circumstances of the wreck and its resulting oil spill (Barnes, 1991: 217). Serious monitoring and publication of violations of the Recommendations has fallen to the NGOs and in particular Greenpeace. Some of the reports emerging by this route in the late 1980s portrayed environmentally degrading practices at various bases, involving the unrestricted burning and dumping of rubbish and discharge of PCBs and heavy metals. Such behaviour would be prohibited under the laws of most advanced countries, let alone the ATS Recommendations.[14] Here it is worth recalling the extreme ecological fragility of the Antarctic and the way in which, for example, it is estimated that oil will take up to one hundred times as long to degrade at Antarctic temperatures as it does under temperate conditions.

Under the CCAMLR there are rules on the reporting of data and fishery statistics to the Commission, plus an observation and inspection system. Unfortunately, following the usual ATS practice, the latter relies upon national inspectors who 'shall remain subject to the jurisdiction of the Contracting Party of which they are nationals' (Art. XXIV(c)). Compliance is a matter for the 'flag state' of offenders against the CCAMLR (Arts XXI and XXII). Detailed provisions for inspection were only agreed in 1989 and are described by one national representative as 'not very binding' (Puissochet, 1991: 75). In the absence of a proper central regulator the CCAMLR is also at the mercy of self-interested national Parties for the provision of essential effort and catch data relating to the Antarctic fisheries.

The Madrid Protocol may be represented as an advance in that it requires Parties to inform others of their own compliance measures along with those taken to ensure the compliance of others (Art. 13). Above all, it puts in place a mandatory system of environmental impact assessment. This involves the production of an initial environmental evaluation (IEE) for a proposed activity. If this indicates that the impact will be 'minor or transitory' the activity may proceed, but if not a comprehensive environmental evaluation (CEE) is required which must be made available in draft form for public scrutiny and will be considered by the Committee for Environmental Protection which will advise the ATCM (Annex I, Arts 2–6). The creation of this body implies a limited element of centralization but its powers only extend to the making of recommendations. Preliminary decisions and the evaluation itself are still in national hands and Watts is correct to identify the publicity provisions as crucial, 'In effect, publicity is seen as the best available means of avoiding environmentally harmful activities being undertaken' (1992: 284–285). An important departure from the environmental protection rules devised for CRAMRA is that whereas the latter covered liability for environmental damage the Madrid Protocol does not.

Scientific Activity

The original Antarctic Treaty was occasioned and justified by the need to continue and extend scientific cooperation, and such endeavours have remained central to the regime. Scientific research has always, therefore, had a particular status within the ATS. Scientific work forms the basis of the activity criterion for admission to the 'club' of Antarctic Treaty Consultative Parties (ATCPs). In practical terms the regulatory side of the regime, as well as providing a framework for the conduct of research and international scientific collaboration, is also heavily dependent upon specialized scientific advice. Advice on matters such as the Agreed Measures on the protection of flora and fauna has been provided by the Scientific Committee for Antarctic Research (SCAR). This body, drawn from national academies of science, was a creation of the International Council of Scientific Unions (ICSU) in the context of the 1958 International Geophysical Year. It thus pre-dates the Antarctic Treaty itself, and although lacking a formal connection with the other elements of the regime is nonetheless regarded as an integral part and the body to which the Parties will turn when in need of scientific advice. The transmission of SCAR's views and advice is very much through national Antarctic Committees and ATCP governments to the Consultative Meetings. According to Beeby (1991: 11) the system 'allows for Treaty Governments to have informal access to a wide spectrum of independent scientific advice through the Scientific Unions and the Committees of ICSU' and SCAR scientists may on occasion take initiatives in advising governments through the same channels. SCAR has, thus, provided the scientific basis for the development of the regime's conservation measures. Its 1976 BIOMASS programme (Biological Investigation of Marine Antarctic

Systems and Stocks) was, for example, very closely associated with the thinking behind the CCAMLR. On the other hand it hardly constitutes an independent or activist 'epistemic community'.[15]

The CCAMLR is the only part of the regime to have its own dedicated scientific knowledge generating body – the Scientific Committee. The complexities of developing an ecosystem approach to utilization and conservation, when so little was known about such fundamentals as krill–predator interaction, necessitated a dedicated advisory body. However, the relationship between the Commission and the Scientific Committee has not been an easy one and illustrates the difficulties that soon arise when scientific advice has a direct bearing on resource issues. Originally conceived as a disinterested source of scientific advice to the Commission, the Committee became politically deadlocked with claims from the Soviet Union that members were representatives of their respective national interests (Stokke, 1996: 139). It lacked the authority to conduct independent research and functioned mainly as a clearing house for data reported piecemeal by governments. Furthermore, in ways readily recognizable from the experience of other regimes, scientific uncertainty soon became an excuse for inaction prompted by other, less objective, considerations (Joyner, 1992: 249–250). From a relationship with the Commission, described by one participant as a 'total lack of dialogue' (Vicuna, 1991: 31), the situation appears, by the late 1980s, to have improved somewhat, with the Commission formally requesting and receiving advice on alternative management strategies. CCAMLR's ambitious ecosystem approach to conservation is still dependent upon scientifically uncertain assumptions about the interaction between the marine species of the Antarctic. The development of that rare commodity, authoritative scientific advice which is accepted as such by members, is regarded as the key to achievement of the aims of the Convention.[16]

Summary

The Antarctic Treaty System regime is of long standing. Created in the depths of the Cold War it secured the demilitarization of the continent and provided a framework within which cooperative scientific endeavour could proceed. As the years have gone by the full significance of such work to an understanding of global physical systems, which was barely perceived in 1959, has become apparent. Above all, the felicitous treatment of the issue of sovereignty and the various territorial claims has ensured that the continent and surrounding ocean retain the status of commons.

As an institutional model of commons governance, the regime has a very low level of formal organization and articulation. Apart from the CCAMLR there is no established secretariat and in relation to all the rules of the regime there is a complete reliance on national inspection and enforcement and an absence of any central authority. This contrasts with the range and specificity of rules which have evolved incrementally over more than 30 years, culminating in the Madrid Protocol. The fact that the regime is restricted to those who can demonstrate an active interest and has functioned as a rather exclusive club, actively resisting all attempts to draw it in to the universal UN system and boycotting General Assembly voting on Antarctica, is surely significant. (There is a distinction here between this and the active relationship between elements of the ATS and specific UN agencies and programmes such as WMO or UNEP.) For one thing, it has ensured that the idea of the continent as a 'common heritage of mankind' with all that implies for its resources and governance and which one might logically expect to apply, has made no headway within the ATS.

Although much of the detailed rule-making has had an incremental character it is possible to observe shifts at the level of norms and principles sufficient to make out a case for the occurrence of regime change. Certainly, the transition from a situation where no mention was made in the 1959 Treaty of environmental norms and principles to their current salience within the regime must constitute 'change'. Rather more interesting are recent alterations in the balance between resource extraction and conservation principles. The consideration of Antarctic resources was initially dominated by the needs of harvesters or potential harvesters – albeit with a greater or lesser degree of concern with the maintenance of stocks. The issues were the familiar ones of the commons literature – the collective action problem posed by open-access resources and the avoidance of 'tragedy' in terms of the collapse of stocks. Antarctic waters had already been the site of such a sequence of events with the self-inflicted death of the pelagic whaling industry. The CCAMLR as a developed common property resource regime can be seen as resting on the principle that the onus is on conservationists to demonstrate that resources should not be extracted. As Joyner (1992: 249) argues: 'This task is impracticable in a fishing regime where decision-making, enforcement and dispute resolution remains biased towards the interests of fishing nations'. The much debated minerals regime appeared to be based on the same principle; mining was allowed *per se* and it was a matter of evaluating and coping with the consequences. Elsewhere in the system, where potentially valuable resources were not at stake, it had been possible to take a much more rigorous stand on the primacy of conservation. However, the significant change that appears to have taken place in the rejection of the CRAMRA and the negotiation of the Madrid Protocol can be portrayed as a rejection of the principle of managed resource exploitation in favour of complete prohibition and 'wilderness values'. Such a conclusion is highly tentative and must always be tempered by a sober awareness that if there was any immediate likelihood of economic exploitation of Antarctic mineral resources the situation would, in all probability, be rather different.

Notes

1. Although it may be noted that Australia has made such a claim.
2. This is specified in Art. 1 of the CCAMLR (1982) as the area 'south of the 60 degrees South latitude and to the Antarctic marine living resources of the area between that latitude and the Antarctic Convergence which form part of the Antarctic marine ecosystem'.
3. For a discussion of possible legal arrangements for the ownership and exploitation of icebergs, including a 'common heritage' regime, see Joyner (1992: Ch. 6).
4. The politics of this process are chronicled by Lorraine Elliott (1994). Of particular importance is her well informed and readable account of the negotiation of CRAMRA, its collapse, and the creation of the Madrid Protocol (Chs 6–8).
5. See for example Jørgensen-Dahl & Østreng (1991: 1) who in their editorial introduction refer to the ATS as a regime and the CCAMLR and CRAMRA as 'these regimes'.
6. Australia, New Zealand, France and Norway all have undisputed claims which they mutually recognize. No other nation extends recognition, except the United Kingdom which recognizes the first two. The UK, Chilean and Argentinean claims overlap and are disputed.
7. For an account of NGO positions and the world park concept see Barnes (1991: 186–221). Suter (1991: Ch. 12) proposes the similar concept of the 'Public Heritage of Humankind'. This not only means that an area, such as Antarctica, should be publicly owned and held in trust for the common benefit, but that there should be a total ban on any resource extraction justified by the inherent importance of the region.
8. I am indebted to David Scrivener for pointing this out.
9. The expectation is that this situation cannot continue and there was a proposal for a secretariat to be located in Argentina, which was vetoed by the UK (Beck & Dodds, 1998: 5).
10. This provision is contained in each of the four annexes to the Protocol, see for example Annex I on Impact Assessment, Art. 8.
11. Suter (1991: 100–110) provides details of infractions involving such things as oil spillage and blasting, the latter having the effect of disturbing bird habitats.
12. For estimates see Drewry (1987), Triggs (1987: 182–183) and Larminie (1991: 91). Larminie writes 'what then is the energy and resource potential of Antarctica? The answer is "don't know but probably poor"'. Some iron ore and coal deposits have been located, but much of the discussion of possible mineral riches is predicated on the Gondwanaland thesis. This postulates a southern super-continent which once joined Antarctica to the other continents of the Southern Hemisphere. Even if potential mineral wealth were shown to exist the difficulties of extraction through the thick ice cap and the logistical problems would appear prohibitive – at least under what would be considered nowadays as reasonable economic assumptions.
13. The Annexes cover: I Environmental Impact Assessment, II Conservation of Flora and Fauna, III Waste Disposal and Management, IV Prevention of Marine Pollution. Annex V on Protected Areas was adopted subsequently at the 16th ATCM in November 1991.
14. See Suter (1991: 108–110) for a summary of NGO reports on waste disposal and oil spills.
15. It is described as by Herr (1996: 97–98) as 'the principal mechanism of the Antarctic epistemic community: that is recognised as authoritative both by the scientific community from which its membership is drawn and by the state

membership of the ATS for its control of professional knowledge'. For Elliott (1994: 195) SCAR did not operate as an epistemic community during the arguments about CRAMRA either in terms of policy advocacy or supporting its own interests. Instead the role of SCAR scientists was diminished in comparison to the influence brought to bear by the NGOs.

16. Stokke (1996: 137–138) observes that there has been a 'steadily improving record of information gathering and scientific study' but that the CCAMLR has 'failed as yet to produce the knowledge necessary for fine-tuned ecosystem management' (Stokke, 1996: 141). For a political description of the flawed development of the Convention see Elliott (1994: 92–101).

5

Outer Space

On 4 October 1957 the Soviet Union launched the first earth-orbiting satellite, Sputnik 1, and for three weeks this tiny object swung around the earth emitting its characteristic bleeping signal. After some embarrassing failures and delays the United States replied with its own Explorer I satellite, launched in 1958. Such primitive beginnings were only precursors to a spectacular technological competition between the then two superpowers for ascendancy on what Americans liked to call the 'high frontier'. The military competition in space and the race to the Moon, culminating in the triumphant landing of Apollo 11 in 1969, gripped the imagination but should not be allowed to obscure the real and growing importance of the utilization of outer space for a host of more mundane terrestrial activities. In this respect the launch of the first commercial communications satellite, Intelsat I (or 'Early Bird') in 1965 is an event of equal, if not greater, significance. It had a launch mass of 68 kg, and its 204 circuits carried telephone calls across the Atlantic for three years. Some indication of the scale of technological progress is provided by comparing the dimensions of a modern Intelsat VI satellite, weighing 3500 kg and with a 33 000 circuit capacity (Gallagher, 1989: 31). Scores of similar satellites in geostationary orbit (GSO) now provide much of the 'global nervous system' of near-instantaneous communication that is almost taken for granted. For decades satellites have been used to transmit television pictures and in the early 1990s direct broadcasting by satellite (DBS), long anticipated and argued over, had become a widespread commercial reality. Remote sensing satellites of increasing power and 'resolution' have allowed the most intensive and revealing study of the earth's surface and environment.

These civilian activities were inevitably accompanied and usually preceded by military applications in reconnaissance and communications – the 'C^3I' without which modern military systems are relatively blind and impotent. It was, for example, estimated in the mid-1980s that some 70% of American C^3I 'assets' were satellite based (OTA, 1986: 42). The 1991 Gulf War provides a case study of some of the implications of all this. In what the US Department of Defense described as the 'first space war' virtually every aspect of Operation Desert Storm, from ground to ground communications between fighting units to identification and tracking the launches of Iraqi Scud missiles, relied upon a vast array of space-based capabilities. At the same time the pioneering CNN TV network, using portable satellite links, provided 'real time' coverage of military action with its live reportage of the attack on Baghdad.[1]

Satellite communications (satcoms) and broadcasting have a number of advantages, even though they are now being challenged by the new generation of broadband fibre optic cable. Satcoms links can be secure and dedicated to particular users. Both telecommunications and broadcast signals can cross national boundaries without using terrestrial links. This raises a number of problems for state governments wishing to control the content of messages crossing their frontiers. Much the same can be said for space-based surveillance systems which also intrude at will into national territory. Foreign corporations may come to possess information on national resources, not available or only available at a price to the national government or local businesses.

Probably the most significant potential role for satcoms is in the development process. Unlike cabled communications, costs are insensitive to distance. Since the Indian satellite experiments of the 1970s it has been clear that access to satellite technology offers the prospect of efficient and cost-effective development of a modern national telecommunications network. As the International Telecommunicatin Union's (ITU's) Maitland Commission (1984) demonstrated, the development potential is enormous. Deployment of satellite technology could begin to address the kind of gross inequities, identified by Maitland, whereby the city of Tokyo has more telephones than the entire African continent.

As far as the information-rich societies of the Northern Hemisphere are concerned, the future of satellite communications appears to be with the new mobile applications. Ships, aircraft and even long-distance trucks are equipped with miniaturized terminals (Inmarsat is the main provider) which allow mobile communications and pinpoint navigational accuracy. By the late 1990s a new generation of systems, allowing personal cellular telephony on a world-wide scale, had begun to be available. They operate through constellations of low earth orbit (LEO) satellites which are accessed directly by mobile phones. This enables subscribers using such satellite phones to communicate with each other from almost anywhere on earth, although it is worth noting that calls are still routed through terrestrial gateways. The main systems are Motorola's Iridium with a constellation of 66 satellites, Globalstar with 48 and Teledesic with no less than 288. Costs are still substantial but will no doubt reduce as the technology develops and the rival commercial systems compete for the international business market. As the global cellphone pioneers have also pointed out their product has major potential for developing countries which lack an orthodox telecommunications infrastructure. The new mobile satcoms could become 'be the great equaliser between richer and poorer' (Lundberg, 1993: 12). However, on past experience this is anything but a foregone conclusion.

These uses of the space commons depend on a particular set of satellite orbits as described in Figure 5.1. The orbit into which a satellite is injected will be determined by its function. Low earth orbits (LEOs) at altitudes of 200–5000 km are usually polar or highly inclined to provide extensive coverage of the earth's surface. They are utilized for a wide range of purposes, photographic reconnaissance and other forms of 'remote sensing', radar, signals intelligence, navigation, meteorology and now some of the new mobile satcoms systems. Geostationary orbit (GSO) is the indispensable basis of current satcoms and broadcasting. It has the unique characteristic of holding a satellite stationary in relation to points on the earth's surface. A satellite launched into this circular and equatorial orbit will have an altitude of approximately 36 000 km and travel in the same direction as the earth's rotation. This means that permanent contact can be maintained with earth-bound transmitters and receivers, whether a massive Intelsat national tele-communications 'gateway' or the humble domestic 'wok on the wall' DBS TV antenna. Additionally, near global coverage is possible through the deploy-ment of just three GSO satellites. (High latitudes would be excluded by the curvature of the earth's surface. To deal with this problem Russian commu-nication satellites (comsats) are launched into a specialized, highly elliptical 'Molniya' orbit.)

Caldwell (1990) has identified outer space and the electromagnetic environment as separate commons. For present purposes it is important to consider them together as the orbit/spectrum resource (or OSR) because the control and utilization of satellites depends upon the use of radio frequencies and, as will be seen, it is the availability and width of frequencies that is a key determinant of the allotment of orbital positions in GSO and indeed the development of the new mobile applications.

In many ways outer space may be regarded as equivalent to the high seas. However, there is a very significant difference in terms of access. For a long time only two countries, the USSR (which up until the end of the 1980s had the largest of all space programmes) and the United States, possessed the necessary launcher technology (either expendable launch vehicles (ELVs), which relied on the same technology as the ICBMs of the Cold War, or the first generation of reusable vehicles such as the US Shuttle or the Soviet Buran). The pioneers have now been joined by the Europeans with their Ariane ELVs and by the Chinese whose 'Long March' launchers are available for commercial hire. Japan, India, Indonesia and Brazil have active space programmes and countries like Israel have a limited capacity to launch their own reconnaissance satellites. There is even more concentration in the very

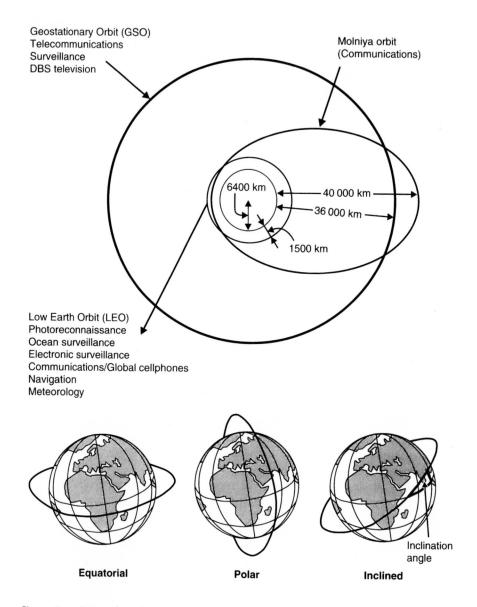

Figure 5.1   Types of satellite orbit

high technology business of building sophisticated satellites, where the market is dominated by North American or European corporations such as Hughes Aerospace, or Matra–Marconi. An interesting, but politically and commercially problematic, trend is the commercial marriage between relatively cheap Chinese or Russian launchers and high technology Western satellites. Asiasat[2] provides a fascinating example, bought second-hand after being rescued from orbit, and re-launched atop a Chinese 'Long March' ELV. It carries the STAR (Satellite TV Asia Region) service.

It is important to stress that the overwhelming majority of countries cannot realistically expect to have their own dedicated national space capability. Launch capacity may be purchased and one of the manufacturing combines will provide the necessary satellite technology, at a price often running into hundreds of million of dollars. For most countries and corporations 'access to orbit' is provided through a range of common user organizations. The most important of these remains Intelsat, a remarkable hybrid containing elements of both international organization and private corporation, that has since 1965 provided the infrastructure of global satellite-based telecommunications. The maritime and mobile equivalent is Inmarsat, but neither now possesses the monopoly that they once enjoyed in relation to private sector space enterprises. There are also regional satellite organizations such as Arabsat and Eutelsat and in some instances, as represented by the Indonesian Palapa, national systems may acquire international users. Nonetheless, the use of outer space, the occupancy of orbital positions and access to the latest technology is very heavily skewed towards the developed North (Hudson, 1985). It is this that fuelled political controversy over the common heritage status of space, 'prior consent' and equitable access to orbit. Here space issues were to form a significant dimension of a broader campaign, launched at UNESCO in 1976, for a New World Information and Communications Order (NWICO), to match the New International Economic Order.

The Space Commons and Space Law

Outer space, including the Moon and the various orbits and related radio frequencies, is now generally regarded as having the properties and status of a commons (Schauer, 1977; Soroos, 1982; Laver, 1984, 1986). The Moon and other celestial bodies are capable of appropriation, but, like Antarctica, have been given commons status in international law. Space itself and the various orbits and frequencies cannot be owned in the same way as a piece of territory or for that matter a 'heavenly body'. There was an ill-starred and rather unscientific attempt by a group of equatorial countries (Bogota Declaration 1976) to claim sovereignty over the GSO as an overhead extension of their national territory. Understandably, this never received wider international support. A continuing legal problem exists at the margin of outer space. This concerns the exact delimitation of the point at which airspace under national jurisdiction becomes the space commons. The United Nations body charged with the development of space law, the Committee on the Peaceful Uses of Outer Space (COPUOS), has been grappling with this problem since its inception in 1959 and it continues to be a perennial item on the agenda of its legal subcommittee. The upper limit of conventional flight is around 25 miles (40 km), while the lowest practical satellite orbit would have a perigee of approximately 100 miles. Between these altitudes the exact delimitation of space and airspace, and indeed the requirement to have any such limit in the first place, are still at issue. The question is of more than purely academic interest because the next generation of space vehicles, already under development in the United States, Europe, India and Japan, will have 'air breathing' engines and be capable of both normal atmospheric and space flight.[3]

As with the Law of the Sea, it is possible to speak of a general legal regime for outer space which lays down important norms and principles and even some rather specific rules. The fundamental document is the 1967 Treaty on Principles Governing the Activities of States in the Exploration and Use of Outer Space Including the Moon and Other Celestial Bodies (hereinafter known as the Outer Space Treaty). Its first two articles merit full quotation:

> Art. I   The exploration and use of outer space, including the moon and other celestial bodies, shall be carried out for the benefit and in the interests of all countries, irrespective of their degree of economic or scientific development, and shall be the province of all mankind.
> Outer space, including the moon and other celestial bodies, shall be free for exploration and use by all without discrimination of any kind, on a basis of equality and in accordance with international law there shall be free access to all areas of celestial bodies.
> Art. II   Outer space, including the moon and other celestial bodies, is not subject to national appropriation by claim of sovereignty, by means of use or occupation or by any other means.

While guaranteeing these fundamentals, the Treaty specifically avoids describing outer space as the common heritage of mankind and uses instead the expression 'province of all mankind'. A subsequent treaty, the 1979 Agreement on the Moon and Other Celestial Bodies does make it explicit that the 'moon and its natural resources are the common heritage of mankind' (Art. 11.1). Furthermore, an international regime to govern their exploitation shall be established (Art. 11.5) and there shall be an 'equitable sharing' of the benefits derived amongst all state parties (Art. 11.7(d)). The parallel with the seabed was clear, although the prospect of commercial resource extraction was even more remote. Another direct correspondence was that the powers that had the technical ability to reach the Moon did not adhere to the agreement.

Although the Carter administration signed the Treaty, ratification was over-taken by some of the same arguments and actors involved in opposition to Part XI of the LoS Convention. The Reagan administration withdrew it from consideration by the Senate and in 1999 it had been ratified by only six states.[4]

A range of specific legal agreements, painstakingly negotiated under the consensus rules of COPUOS, develop other aspects of the 1967 Treaty. They include the 1968 Convention on the Rescue of Astronauts; the 1972 Convention on liability for damage caused by space objects; and the 1975 Convention on the registration of objects launched into space. The latter requires signatories to submit basic information on their launches to the United Nations.

This corpus of space law may represent a regime in terms of legal usage, but not in terms of the idea that regimes govern issue areas. It is worth recalling Keohane's (1984) description of the latter as 'sets of issues that are in fact dealt with in common negotiations and by the same or closely coordinated bureaucracies'. Ploman (1984: 155) was undoubtedly correct when he wrote:

> this concept of regimes is helpful in revealing the absence of any single overall regime. Instead there has emerged a number of partly overlapping, often uncoordinated and sometimes contradictory or competing regimes. Furthermore, satellite communication is subject to the influence or application of regimes evolving in other areas which directly or indirectly relate or are made to relate to satellite regimes.

Establishing the relevant issue areas for the space commons is inevitably a subjective business, but the following four seem justifiable in relation to Keohane's criterion:

- Military uses of space – ASATs.
- Space debris and the environment.
- Information flow – 'prior consent'.
- The orbit spectrum resource (OSR).

Throughout the Cold War the use of space was intimately related to the superpower confrontation. It posed such evident risks in terms of the possible deployment of weapons of mass destruction in space and the more imminent prospect of destabilizing interference with space-based warning, communications and command assets, that military issues appeared of pre-eminent importance. In the early 1980s these concerns were heightened by the awareness that many components of the Strategic Defense Initiative (SDI), the attempt to construct a defence against ICBM attack and hence to challenge the established bases of mutual deterrence, would be located in orbit. By the final years of the Second Cold War these were very high profile issues and led to a real academic and political concern with 'war in space' (Jasani & Lee, 1984; Kirby & Robson, 1987; Lee, 1987).

A closely connected set of concerns grew up around the dangers of radioactive materials in orbit and the environmental implications of the use of space. The environment of outer space itself was not at issue. More recently, the physical dangers to current and future space flight arising from the pollution of the space commons, much of it deriving from military activities, has been considered. This has particular relevance to the safety of the permanent and inhabited space stations now under development.

The commercial implications of the utilization of space have been inherently transnational. DBS transmission 'footprints' spill across frontiers, remote sensing does not respect the sanctity of sovereign territory and satellites support extensive transborder data flows. The new generation of small and mobile applications have the potential to operate completely outside national networks and receiving antenna (Motorola built a technically redundant provision to route calls through national gateways in their Iridium system, simply to assuage the fears of national governments). As a result of these technical developments a range of information flow issues have been significant within COPUOS, UNESCO and elsewhere.

The most fundamental of all the issue areas covers the occupancy and use of the orbital positions (particularly in GSO) and frequencies upon which both civil and military use of the space commons depend. They have been treated as extension of the management of terrestrial frequency use and thus fall within the ambit of the ITU. There are questions here of equitable access, efficiency and coordination in the use of a scarce resource which raise classic commons problems in a new setting.

Military Uses

Although there have been numerous attempts to establish the demilitarization of the space commons and the 1967 Treaty urges peaceful use and exploration (banning under Art. 4 the orbiting of mass destruction weapons), there is no likelihood that military activity can be prohibited as it is in Antarctica. The situation is more akin to the Law of the Sea, where great powers have always ensured their freedom to deploy and use naval forces. It will be recalled that maintaining naval rights of passage and access was a shared consideration for both the US and USSR during the UNCLOS III. The situation in space may be regarded as being essentially similar. There are a number of specific rules and prohibitions, equivalent to those that constrain the use of force at sea. Most of them emerged not from multilateral fora but from bilateral negotiations between the two old superpowers, which were effectively the only two space powers. Typically, also, they constituted part of the fabric of terrestrial arms control. Thus the 1963 Atmospheric Test Ban Treaty prohibits nuclear detonations in outer space and the 1972 SALT I agreement (Art. 5) prohibits testing and deployment of anti-ballistic missile systems (ABMs) in space, something that was to be subject to highly controversial reinterpretation by advocates of the SDI programme. Similarly, the unratified SALT II Treaty contained provisions that bound the parties not to interfere with each other's 'national technical means' of verification – a euphemism for reconnaissance satellites.

During the 1980s the most significant area of potential military space activity not covered by a formal ratified agreement (and one intimately related to SDI) was the testing and deployment of anti-satellite weapons (ASATs). Both the US and USSR had been experimenting with such potential weapons since the 1960s. Between 1968 and 1982 the USSR staged at least 20 tests of a co-orbital interceptor whereby an explosive satellite was manoeuvered into an orbit alongside the target. Subsequently, the Soviet Government declared a unilateral testing moratorium and introduced a draft treaty at the UN outlawing ASAT weapons. Having abandoned early tests in 1975, the Americans proceeded to develop a 'direct ascent' weapon where a homing missile was launched at a LEO satellite from a high flying F15 fighter. It had one successful test in 1985 whereupon the US Congress imposed its own restraint by refusing to fund further tests during 1986/87.

There had been direct bilateral talks on ASAT arms control, initiated in 1978, but these were terminated by the onset of the Second Cold War in 1979. Neither did Soviet inspired moves at the UN amount to anything in terms of the negotiation of a formal set of rules. Nonetheless, it is possible to argue that an ASAT regime grew up on the basis of largely implicit understandings between the two major space protagonists. In Bellany's (1987: 230–231) view 'The evidence for . . . a de facto or tacit piece of arms control amounting to a reciprocated hesitation to acquire a military significant antisatellite capability – is quite impressive'. Characterizing it as a regime requires a certain amount of supposition, but it would be based upon a principle, apparently shared since the early 1960s, that 'overhead surveillance' by satellite was a legitimate activity. The protection of monitoring facilities agreed in SALT II would have both reflected and strengthened this. Also, self-proclaimed 'scientific' space observation programmes may have been significant 'in getting the people of earth used to having their photographs taken from space, a gradual development that bore upon the legitimation of satellite reconnaissance, however indirectly' (Burrows, 1986: 142). Whether there was also a principle that space constituted a sanctuary is much more problematic. Gray (1983) argued forcefully that in time of war between 'space powers' and in the context of the critical role of space-based systems it would be idle to suggest that the norms associated with space as a sanctuary would ever be upheld. On the other hand restraint, at least in peace time, which amounted to a shared norm that satellites should not be interfered with and an understanding that ASAT weapons would not be deployed even though a capability was retained, did make military sense. This was because, as Bellany (1987) and Stares (1987) have demonstrated, the military users of space shared a high level of mutual vulnerability in terms of their dependence on fragile orbital 'assets'.

The very close monitoring by each space power of the other's orbital activities ensured that attempts to test ASAT systems were well known and that each side would have been aware of a move by the other towards an operational capability. However, as the extensive debates of the 1980s about space arms control demonstrated, the establishment of precise rules and limitations allied to reliable verification threw up a range of difficulties.[5] For one thing there were problems in distinguishing effectively between ASAT capabilities and the components of ballistic missile defence systems (SDI), and the USSR would have had an interest in tying both together in a blanket ban, unacceptable to the US, on all space weapons. Arguably, this was an instance in which real constraints were better achieved by an informal understanding than by the attempt to negotiate a full ASAT treaty. General negotiations on space de-militarization have continued in the UN Disarmament Committee's Ad Hoc Committee on the Prevention of an Arms Race in Outer Space, but without significant result.

The ending of the Cold War, the downgrading of the SDI programme and the relative eclipse of Russia as a space power, make the fierce concerns of the 1980s seem very dated. The US military position in space is, for the moment, unchallenged. There is no better illustration than the 1991 Gulf War. Iraq was denied the minimal satellite intelligence that would have been sufficient to alert its forces to the massive feint that was the key to allied military strategy. It had no space assets of its own, no way of interfering with allied military uses of space and was denied access to commercial sources of intelligence, such as the French SPOT reconnaissance satellite, by a UN embargo.[6] Such conditions are unlikely to be repeated as more countries assimilate this military experience and acquire space capabilities. Similarly, the proliferation of 'civilian' remote sensing satellites will make it increasingly difficult to deny 'overhead intelligence' to potential adversaries. American military hegemony over the space commons is, thus, unlikely to be permanent. This much was recognized by the departing US Vice President Quayle who, in his last act as Chairman of the National Space Committee, observed that maintaining the supremacy demonstrated in the Gulf War would require the US to 'develop a comprehensive anti-satellite capability to deny the military use of space to future enemies'.[7]

Environment and Space Debris

When the environmental dimensions of space activity were first considered, concern was exclusively and understandably focused upon the terrestrial implications. A primary issue was liability for damage caused by the remains of space vehicles that survived the rigours of re-entry into the earth's atmosphere. A well known case, which tested the rules, was provided by the fate of Kosmos 954, a Soviet military ocean reconnaissance satellite utilizing a nuclear power plant to drive its sensors. The planned manoeuvre to place the used reactor into a 'safer' higher orbit failed and in early 1978 it re-entered, scattering radioactive debris across a stretch of uninhabited Canadian tundra.[8] Subsequent events demonstrated the operation of the existing international law on liability (1972 Space Objects Liability Convention, based on Art. 7 of the 1967 Treaty) when the Soviet Government paid compensation. The question of the regulation of nuclear power sources on board space vehicles has been considered by COPUOS which produced a set of regulatory 'principles' in 1995.[9] The other original concern, as dramatized in Michael Crichton's 1969 novel *The Andromeda Strain*, was the danger of terrestrial contamination by returning space vehicles and astronauts. This is covered under Art. 9 of the 1967 Treaty and NASA, at least, took elaborate quarantine precautions with its Apollo Program astronauts.[10]

The environment of outer space itself was not covered under these provisions. It has recently become an issue of some significance because of the dangers posed by the accumulation of space debris. The detritus of over 40 years of satellite launches and often reckless military experimentation now clutter the space commons. Approximately 9000 objects greater than 10 cm in size, of which only around 5% are functional spacecraft, have been observed and catalogued in near-earth orbit (Wilkinson *et al.*, 1999: 30). There are also hundreds of thousands of smaller objects, some the product of human activity but others comprising cosmic dust and micrometeorites. The sources of manufactured space debris are various:

> defunct satellites, discarded rocket stages, hardware released during launch and on-orbit operation, flakes of degraded surface coatings, lumps of unburnt solid rocket fuel, droplets of coolant from leaking nuclear power systems and fragments from collision and on-board explosion (the latter being the greatest source of catalogued debris) (Wilkinson *et al.*, 1999: 28).

The main risk of collision and loss or degradation of spacecraft occurs in LEO. In such orbits debris will impact spacecraft at an average velocity of 12 km per second with even relatively small objects having a devastating effect. At the same time satellite population density is much higher in LEO than in GSO (about 100 satellites are launched into LEO per annum compared with around 20 injected into GSO). LEO usage is bound to increase over the next decades, especially with the introduction of the large satellite constellations associated with the new generation of global cellphones. There has long been evidence of debris impact damage. Each Shuttle mission for example registered 30–40 small impacts and windscreens have had to be replaced. There were also suspicions that otherwise unexplained satellite failures were attributable to collision with space debris, but it was only in 1996 that the first fully authenticated case was confirmed.[11] Modelling of the future debris environment predicts an increasing risk of satellite failure over the next 50 years if measures to address the problem are not adopted. In the worst case there is the possibility of 'collision cascading' in which collisions generate debris which initiate further collisions leading to self-sustaining and 'uncontrollable population growth' (Wilkinson *et al.*, 1999: 39).

In GSO the problem is less severe because of lower population density and impact velocities. Nonetheless, by the end of the 1980s there were already 'about 100 large objects, particularly the Soviet Ekran TV satellites and Proton upper stages, weaving around the geostationary orbit arc' (Chenard, 1990: 255). Debris in GSO has a much longer life-time than LEO simply because there is very little orbital peturbation to remove defunct satellites and other objects from the vicinity of operational spacecraft. LEO, however, possesses what amounts to a natural 'sink'. This derives from the phenomenon of 'atmospheric drag' where objects in LEO will eventually be pulled back into the earth's atmosphere.

There are various ways of mitigating the space debris problem. Spacecraft design and construction can be altered to improve protection against small pieces of debris and in the case of the US Shuttle orbiter by including evasion manoeuvres in flight planning. It also makes sense to encourage space users to cooperate in reducing the extent of the debris problem by adopting improved practices. This, in essence, is the basis of what may be described as an embryonic space debris regime. Given the uncertainty as to the extent of the hazard it is in accord with the 'precautionary principle' and includes the fundamental norm that space users have a responsibility to minimize the amount of debris arising from their operations. There is also an understanding that maximum transparency is desirable in terms of information as to the extent of the problem and best practice in terms of mitigation measures. The intention of those involved in developing the regime is to get the best advice available as quickly and transparently as possible to the space industry so that there will be no excuse for newcomers to adopt practices giving rise to unnecessary pollution of the space environment.[12]

Unlike the other environmental regimes covered in this book, a space debris regime is not yet formally codified but rests upon a number of understandings that serve to coordinate behaviour amongst the relatively restricted group of spacefaring nations, spacecraft operators and manufacturers. Government space agencies have been aware of the potentially damaging effects of space debris for some years, but Cold War space competition between the USA and USSR and the lack of an internationally agreed estimate of the extent of the problem impeded progress both at COPUOS and elsewhere. An organizational framework for collaborative action only emerged in 1993 with the creation of the Inter-Agency Orbital Debris Coordination Committee (IADC). This body, which includes the main national space agencies and the European Space Agency, enables members to exchange information on space debris activities, cooperate in research and identify debris mitigation options. The main thrust of its initial work has been to promote a collective understanding of the problem by persuading members to correlate the data from their respective national research programmes. Such research is needed in order to justify the sets of standards for debris mitigation being developed by IADC to which all new space programmes are urged to adhere.

Informal rules and understandings on mitigation were already in evidence in the late 1980s. One such rule would be the avoidance of the previous practice of detonating 'sensitive' reconnaissance satellites once they had reached the end of their useful life. During the 1980s officials at the Johnson Space Center had regular discussions with their Soviet Glavkosmos counterparts on orbital debris issues. According to one US participant, 'We talked to the Soviets on the detonation problem, but they took action before any formal interchange' (Chenard, 1990: 254). Other rules include the alignment of national practices on the 'passivation' of the upper stages of launchers and the de-orbiting or re-orbiting of spent satellites. The need to 'passivate' upper stages arises from the propensity of unused liquid propellants to explode after separation from the satellite. It involves ensuring that all remaining fuel is either vented or burned. NASA required these techniques from 1982, followed by the European Space Agency in its Ariane launches from 1989.

De-orbiting or re-orbiting spent satellites is now accepted practice, even though it will impose significant costs in terms of a reduction in the amount of fuel that can be carried for normal station-keeping. (Satellites continuously use small amounts of fuel to power the thrusters which make the small course corrections required to hold an orbital position. Re-boosting a satellite into a different orbit or out of orbit requires a much larger amount of fuel.) There is, thus, a trade-off between de- or re-orbiting and the operational lifetime of the satellite.[13] In GSO the practice, now followed by major operators (Intelsat, Eutelsat, Telesat Canada, ESA, Aussat, the Indian ISRO and all the US agencies) is to re-boost spent satellites into a higher 'junkyard orbit'. This relates to the wider question of evolving rules for orbital 'good house-keeping' which are the province of the ITU. The latter also has licensing rules for operators, requiring them to consider and report upon debris mitigation practices (Wilkinson et al., 1999: 43). Other approaches include standards for spacecraft construction, particularly involving non-flaking paint materials and new types of battery which, unlike current nickel–cadmium varieties, are not liable to explode.

The above 'standards and recommended practices' are not binding in international law and they will involve additional costs which operators may be unwilling to pay. Furthermore, although there are liability clauses in the

Outer Space Treaty (and a supplementary Liability Convention), they are not applicable to damage to environment of the space commons (Gorove & Kamenetskaya, 1995: 486). It is, as yet, impossible to establish, for insurance or other purposes, the responsibility for particular pieces of debris that may cause damage. Nonetheless, the IADC has demonstrated that it is possible to track and catalogue debris in excess of 10 cm and, importantly, to agree the data. Detonations and failure to de-orbit spacecraft will be observed and publicized in space industry circles, which provides some incentive to adhere to the rules of good orbital practice. While it is impossible to impose international standards and regulations it is possible to involve spacefaring nations in international technical collaboration and to get them to institute mitigation measures because 'they know it is the right thing to do'.[14]

The formal codification of a regime for the outer space environment is recognized to be within the remit of COPUOS. However, there was initial resistance from the United States, France and Russia to developing country pressure for this UN body to take up the question of orbital debris. The Scientific and Technical Sub-Committee of COPUOS undertook a multi-year work programme on the problem from 1996 to 1998 informed by the findings of IADC and covering agreed models, measurement and mitigation procedures. Its legal counterpart did not commence the drafting of text for a binding agreement although there were suggestions that, for example, a 'launcher pays' principle in respect of damage to the space environment should be instituted.[15] According to Gorove & Kamenetskaya (1995: 487): 'There can be no question that the lacunae in international space law for the protection of the "space commons" must be addressed, either specifically through a new space treaty or . . . through international environmental law'. The portents for this are not good, for in the related case of nuclear power sources in space it required the shock of the Kosmos 952 accident and 10 years of deliberation in the COPUOS technical and legal sub-committees to produce the 1995 legal principles. In the meantime it is important that the regime based upon technical collaboration and understanding between the national agencies continues to develop.

Information Flow

The revolution in information and communications technologies has been central to debates about globalization and indeed the continued viability of sovereign states as the principal units of the international system. The transnational gathering and dissemination of information using electronic media (this would include data flows (TDF), broadcasting and remote sensing activities) has long been a significant issue in North–South relations. During the 1970s and early 1980s the South's campaign for NWICO, a New World Information and Communications Order, dominated debate within

UNESCO. In 1979 the UNESCO Mass Media Declaration, supported by many Southern countries and the then Soviet bloc, attempted to erect principles legitimizing control over transnational media in the name of protecting national cultural sovereignty against the 'electronic colonialism' of the North. For the latter this was an attack on cherished liberal principles of freedom of speech and information and contributed much to the American and British decisions to leave UNESCO in 1984 and 1985 (McPhail, 1981; Roach, 1990). Since the NWICO debates, information issues have been recognized as having an ever increasing commercial as well as cultural and political significance as evidenced in the negotiation of rules for 'trade in services' in GATT's Uruguay Round.[16]

Space-based activities are highly relevant to global information flows because, as we have seen, they provide much of the technical infrastructure. However, the aspects of the wider information debate that are specific to the space commons (a good test is their inclusion on COPUOS agendas) are DBS and remote sensing. Both could be regarded as inherently threatening to national cultural and economic sovereignty, particularly in the developing world.

Developed by NASA under pressure from the US Congress, remote sensing of the earth's environment, meteorology and natural resources has enormous implications, most of which are potentially beneficial. There is also a legitimate concern that commercially significant satellite imagery and data relating to a particular country will neither be under the control of its government nor even available to its decision-makers and business people. At first the pioneering US Landsat system attempted to meet such objections by emphasizing research and developmental functions and by supplying imagery freely on the basis of a low flat rate fee. Increasingly Landsat was also used for intelligence purposes and by transnationals surveying for new sources of oil or other raw materials. In 1985 the US Government transferred the operation to the private sector and information from the latest generation of US EOS (Earth Observation System) is priced at commercial rates. EOS is now in competition with the French SPOT (Système pour l'Observation de la Terre) and with the Russian space industry, eager for hard currency.[17] It is easy to understand the resentment in many developing countries at what may be seen as intrusion into national economic sovereignty and the way in which, according to one commentator, 'Absurdly priced information is kept from its main user market by feeble and ineffective attempts at clawing back a profit where none can exist' (Baker, 1989: 54).

Even more controversial has been the development of satellite TV broadcasting and DBS. Originally satellite TV was transmitted (using C band) to large master antennae which then delivered the signal to a cable network. In principle, at least, this allowed national regulation of the cable networks. In practice the explosion of satellite TV channels, using this now primitive technology, has been largely unregulated. India provides a fascinating example of sudden transition from a state broadcasting monopoly to a plethora of foreign-based channels. In 1991 Satellite TV Asia Region (STAR), with its headquarters in Hong Kong, began transmitting five channels including MTV and BBC World TV to over a million homes in India and an estimated 10 million across Asia (Hong Kong, the Philippines, Pakistan, Korea and even Israel are covered by the transmission 'footprints' of the satellite's two beams). STAR's service utilized 10 of the 24 transponders on Asiasat 1 (the Chinese-launched second-hand satellite mentioned above and in note 2). The service was 'free to air', that is to say anyone with an antenna could receive and distribute it (Simon, 1993: 6–7). In 1993 Rupert Murdoch's News International acquired a controlling interest in STAR. Indian consumers can also receive CNN and Pakistan TV by satellite and cable and by 2000 over a third of the population were included in the satellite TV audience.

In Europe and especially in Britain, DBS, which requires no cable links from a master antenna, has become a commercial reality in the shape of Sky TV (also owned by News International). The full extent of the market penetration of DBS and the implications for terrestrial broadcasters and the maintenance of national cultural integrity may still be unclear but they are widely acknowl-edged to be momentous. Such uses of the space commons were hotly argued over long before they became an operational and commercial reality. During the 1970s COPUOS debated the relative merits of 'prior consent' to DBS transmissions originating beyond national boundaries as opposed to the free flow of information. There was a predictable deadlock between advocates of the NWICO who tended, along with the USSR, to demand 'prior consent' and the United States and her allies. Supporters of 'prior consent' proceeded to obtain a 1982 General Assembly Resolution on 'Principles Governing the Use of Artificial Earth Satellites for International Direct Television Broadcasting'. Although passed by a majority, Western nations dissented strongly and made it clear that they were in no way bound by its provisions (Williams, 1985). ITU provided an alternative forum for the development of a DBS regime. The 1977 World Administrative Radio Conference (WARC) drew up a European direct satellite broadcasting plan (known in ITU terms as the plan for the Broadcast Satellite Services). As well as allocating frequencies to member nations it also defined 'coverage areas' and included a statement that states should minimize the spillover of the 'footprints' of their DBS systems into other countries. The intent was clearly to 'maintain the control of national governments over broadcasting and to limit the ability of Western satellites to broadcast directly to Eastern Bloc countries' (Veljanovski, 1987: 37). The outcome provided a classic instance of the way in which regulation may be outstripped by techno-logical change. As the Luxembourg backers of the Astra satellite, that carried Sky DBS transmissions, discovered, it was possible, by taking advantage of technical developments, to broadcast completely outside the restrictions of the plan. By contrast, British Satellite Broadcasting (BSB), the ill-fated rival of Sky, was saddled with the 'official' frequencies allotted to Britain under the ITU plan.[18]

The attempt to erect a regime for remote sensing ran in parallel with the DBS debates at COPUOUS (consideration commenced in 1974) and fared no better. It, too, hinged upon the question of 'prior consent' as opposed to 'free flow'. It was argued that states subject to remote sensing without having granted their prior consent should have legal redress. The United States supported by other advanced industrial countries argued that it was technically impossible to confine sensing within national boundaries and that it was 'primarily a space activity' and therefore subject to the 1967 Treaty and the principle that 'outer space is free for exploration and use by all states' (Magdelenat, 1985: 134). Accordingly, the right both to sense and to disseminate data was demanded.

There is, therefore, either no 'information regime' for the space commons, or a regime that is thoroughly permissive, allowing complete freedom to direct broadcasts across frontiers and to remotely sense the territory of other states. National policies and restrictions exist, but they are essentially uncoordinated at the international level. The countries of the European Union, for example, are concerned to implement policies which safeguard indigenous media industries, but also to deregulate access to and control of satellite broadcasting. Some governments, such as that of Saudi Arabia, have taken draconian measures to protect their cultural sovereignty by banning DBS dishes. Others, like China, have had political irritants like BBC World TV removed from transmissions by courtesy of the owner.[19] In view of the scale of demand, the corporate interests involved, the potency of the technology and the world-wide trend towards deregulation in telecommunications and broadcasting, such policies are likely to be seen in the longer run as having a particularly 'Canute-like' character.

Orbit and Spectrum

Although the potential of the use of geostationary orbit (GSO) had been known since Arthur C. Clarke's pioneering work in the mid-1940s, it was not until the late 1950s that the development of space launchers made it a practical possibility requiring international regulation. As frequency use determines the spacing of GSO satellites, it was natural that the new regime for using this militarily and commercially significant dimension of the space commons should have developed from the existing regime for the regulation of the radio frequency spectrum. This was and is operated by the International Telecommunication Union (ITU), an organization directly descended from the International Telegraphic Union of 1865 and, since the 1930s, the ITU has taken over the work of the old Radiotelegraph Conferences. The Union first considered the regulation of GSO at an extraordinary conference convened in 1963 – some years before the beginnings of the widespread commercial use of the orbit during the 1970s.

The question of orbit/spectrum resource scarcity is a moot point of some political significance. The OSR cannot be considered as a fixed stock of frequencies and orbital positions. Unlike fish stocks or seabed minerals it is infinitely reusable although its utility can be much degraded if use is uncoordinated. At any moment in time its potential will be governed by available technology, which has constantly served to open up new areas to exploitation – higher and higher frequencies and sophisticated spectrum expansion and reuse techniques. In some areas of terrestrial broadcasting, for example short and medium wave, there has long been severe congestion and interference. Scarcity in GSO is not a matter of physical congestion or a 'traffic jam in the heavens'. Rather, it turns upon the occupancy of certain highly favoured positions (for example the North American arc of 85° to 115° W) and very complex engineering questions of the spacing required between satellites in order to avoid interference in the GHz bands of the spectrum, which themselves may be in short supply. The present upper limit for civil GSO satellites is estimated at around 180.[20] The engineering solution is to encourage the use of the Ku band (14/11 GHz) and above, which allows more efficient use of spectrum and closer orbital spacing. During the 1980s the overwhelming majority of established systems used the less sophisticated, lower frequency C band (6/4 GHz). Specialists from the developed world often argue that there is no shortage problem that cannot be solved by using higher frequencies and new technologies. However, these involve higher cost and technical sophistication and would not really be the preferred option for LDC new entrants to the satellite business. As Laver (1986: 368) noted, this meant that 'the OSR, in relation to cheap slots where people want to use them, was already in danger of congestion'.

Principles and Norms

The principles and norms associated with radio frequency spectrum use have evolved over the period since 1903 when international arrangements on their coordinated use of the frequency spectrum were first discussed (at the Berlin Radiotelegraph Conference). The events leading to these pioneering efforts at international radio frequency regulation are of some historical interest. In 1902 Prince Heinrich of Prussia travelled to the United States on board the liner *Kronprinz Wilhelm* – equipped with the new Marconi ship-to-shore radio. This was demonstrated to the enthusiastic prince who resolved, on his return, to use the new technology to send a courtesy message of thanks to his host, President Theodore Roosevelt. Unfortunately, he was embarked on the liner *Deutschland*, which was equipped with German manufactured Slaby Arco Braun radio. The receiving station at Nantucket, controlled by Marconi, refused to take the call on the grounds that a rival firm's equipment was being used. It was at this time Marconi's policy, enthusiastically backed by the British Admiralty, to create a monopoly in ship-to-shore communications. The reactions of the Prince can be imagined and his protests led directly to the calling of the first International Radiotelegraph Conference held at the Berlin Imperial Post Office.[21] This meeting established the principle that the use of the radio frequency spectrum is open to all, regardless of equipment used, and the norm that operators were bound to receive and exchange messages.[22]

Another principle, of long standing, that tends to inform all the work of the Union, is that of the 'rational use' of telecommunications and the encouragement of efficiency in the use of the spectrum. Underlying this is the assumption that ITU, as arguably the world's oldest international organization, has an essentially 'functional' rather than 'political' character. Managing spectrum and orbit is a technical undertaking to which there are rational and consensual engineering solutions. 'Politics' has no proper place in this undertaking. Hence, there was widespread alarm when Third World demands for equity were introduced into deliberations. The campaign for equity in orbit was represented as 'the injection of political ambitions' into the ITU's work and was 'at variance (with) if not outright contradiction of, the basic goals of the ITU' defined as 'maximizing capacity and efficiency of use of the scarce natural resources of the geostationary orbit and spectrum and avoiding harmful interference, as increasing numbers of users emerge' (Stowe, 1983: 63). Associated with this is a discreet avoidance of discussion of the military uses of spectrum and orbit.[23]

Although not explicitly stated within the Conventions, the ITU frequency regime evolved according to the principle that the spectrum was a common pool resource open to exploitation by those who were in an economic and technical position to do so. The ITU managed a limited CPR regime of great technical complexity, designed to coordinate the activities of users of the spectrum such as to avoid mutual interference and ensure common technical standards and interoperability.[24] The rights of users have long been defined in terms of protection from harmful interference once a frequency has been duly registered. Corresponding normative obligations are placed upon newcomers to coordinate with existing assignments. Under the old frequency regime 'squatters' rights' to frequencies were the accepted principle (*a posteriori* rights vesting). The right to notify and assign particular frequencies and orbital slots to actual users rests firmly with national administrations (ITU parlance for state telecommunication authorities). Thus, even large-scale common user organizations such as Intelsat and Inmarsat have to obtain orbital positions and frequencies through national Notifying Administrations. ITU regulatory activity is pervaded by a voluntaristic norm. This can be gauged from a reading of the Conventions, where extensive use is made of the verbs 'coordinate', 'harmonize' and 'promote'. In strict legal terms this is evidenced by the right in ITU negotiations to enter national 'reservations' and 'footnotes' to the various acts and regulations – essentially these are legitimized 'opt out' procedures.

The initial response to the birth of satellite communications was to create a regime which adopted the norms and principles of the existing spectrum arrangements, and was in a sense merely an extension of it.[25] However, the development of principles and norms relating to GSO was not a matter confined to ITU. The resolutions of the UN General Assembly, the terms of the 1967 Outer Space Treaty and the UN Committee on Peaceful Uses (COPUOS), all provided significant cues to reassessment.[26] The Outer Space Treaty introduced not common heritage principles, but the idea that space was certainly *res communis*, 'the province of all mankind' to be used without discrimination and on the basis of equality and free access (Art. 1). It was not to be subject to 'national appropriation by claim of sovereignty, by means of use or occupation or by any other means' (Art. 2).

It took a concerted campaign by new LDC members of the ITU to insert such principles and norms into the emerging GSO regime, even at a largely declaratory level. The 1971 WARC adopted the principle that the registration and use of a satellite orbital position could not be construed as giving a permanent property right and should not create an obstacle to the establishment of space systems by other countries.[27] A more definitive statement was provided in the 1973 ITU Convention.[28] This confirmed the principle that GSO and associated frequencies were a 'limited natural resource' but also established the right of 'equitable access . . . taking into account the special needs of the developing countries'. An attempt was made at the 1979 General WARC, at Indian instigation and supported by an LDC majority at the conference, to give these declarations some operational reality. The relevant resolution (no. 3) enjoined members to 'guarantee in practice for all countries, equitable access to geostationary orbit and the frequency bands allocated to space services'. Because the GSO was a valuable resource, almost monopolized by developed countries, this resolution moved some way towards common heritage ideas. It committed the ITU to the holding of a special conference, not later than 1984, to implement the principle. (This was the origin of the 1985–88 Orbit WARC.) The implication of attempting to implement 'equitable access' was that the allocative principles of the regime would have to be altered towards the *a priori* planning of the common resource. Because the success or failure of this enterprise was so much bound up, not with the elaboration of general principles, but with very detailed rule-making, the outcome of the 1985–88 Orbit WARC will be considered below.

Decision-making Procedures

The organizational architecture of ITU that supports the spectrum/orbit regime is of long-standing.[29] At the highest level the Plenipotentiary Conference of all members, held every four years, sets general directions. In the interim period an Administrative Council, composed of an elected 25% of the membership acts in its stead. The policy function, the making of collective choices about allocations and planning, was carried out by World Administrative Radio Conferences, to which all members are invited. The allocation of the entire radio spectrum to services (including space services) was carried out by general WARCs, usually at 20 year intervals (the last meeting under the old arrangements was held in 1979). In the interim more specialized conferences were held to plan the arrangements for particular services. A similar situation pertained for the the terrestrial telecommunications work of the Union. By the beginning of the 1990s it was clear that such leisurely arrangements could not survive the onslaught of technological change and commercial pressure in the telecommunications sector (Staple *et al.*, 1989; Mohan & Vogler, 1991). Accordingly the ITU was extensively reorganized by the adoption of a new Constitution and Convention in March 1993. This divides the organization's work into three sectors: Standardization (including terrestial telecommunications), Radiocommunication (including spectrum and orbit questions) and Development. Each has its own regular conference, technical study groups and secretariat.

Now within the new Radiocommunication Sector the International Frequency Registration Board (IFRB) supervises the notification, coordination and assignment procedure and keeps the Master Register of frequency assignments. Standard setting is conducted by the International Consultative Committee for Radio (CCIR), which comprises not only representatives of national administrations, but private sector participants and the representatives of Recognized Private Operating Agencies (RPOAs).

At plenipotentiaries and conferences the rule of one state (Administration) one vote prevails. Thus it has been possible, for example at the 1979 General WARC, to mobilize a winning coalition from among developing state members in support of major shifts in declaratory policy (Vogler, 1984). However, in the all important technical working groups that are the setting of detailed negotiation there is an understanding that decisions should be by consensus. Equally if not more significant is the very low rate of participation in such technical work by most developing countries (Codding, 1981).

Rules

The technical standards involved in GSO use are one complex part of an immense edifice that has been constructed by the ITU. Much of this is to be found in the Radio Regulations which are the indispensable guide to frequency use and run to over 1000 densely packed pages (also subject to recent reform). There are also protocols and standards for satellite access (TDMA, SPADE etc.) which have been developed by the CCIR and its various working groups. It is expected, and is in fact overwhelmingly the case, that members and those to whom they assign frequencies and GSO slots behave in conformity with the Regulations and standards.

Following the disputes over principles, noted above, distribution rules have been at the heart of recent controversies over GSO. There are two main types of distribution, allocation and assignment. Allocation denotes the process whereby frequencies are divided up between different services, i.e. the C and Ku GHz bands are divided up to accommodate Broadcast Satellite Services, Fixed Satellite Services etc. (The ITU did this in its General WARCs for the entire frequency spectrum and the list is contained in the Radio Regulations Table of Frequency Allocations.) This is an exercise in collective allocation, but not to individual users. The latter frequently differ over such allocations because they may give higher priorities to different services. For example, there have been conflicts over the relative portions of the HF (short wave) part of the spectrum that should be allocated to broadcasting as opposed to telephony.

The question of who will actually use the frequencies, within the various service bands, is determined by assignment rules and procedures. The existing and dominant principle is, as noted above, *a posteriori* rights vesting in which members can notify an intention to use a GSO slot and frequencies and after going through 'coordination' procedures via the IFRB, have the assignment entered in the Master Frequency Register and thus protected. Such a rule clearly contradicts equal access principles and has been extensively challenged by latecomers to the satellite business. They have argued that it enshrines the privileged position of the advanced users, who are already in occupation of the 'best' orbital positions, and imposes coordination costs on latecomers.

The alternative model is *a priori* planning or the centralized allotment of GSO orbital positions on an equitable basis. There were precedents for such planning exercises within ITU, especially in areas like European terrestrial broadcasting where relatively free access to frequencies had brought overcrowding and electronic chaos (in 1977 a European Direct Satellite Broadcasting Plan was drawn up). Methods and Rules for an equal access allotment plan for fixed satellite services were painstakingly negotiated in a two-session space WARC held in 1985 and 1988 (WARC-ORB). The details are exceedingly complex but may be summarized as follows. A plan was devised, coming into force in 1990, which gave every ITU Administration, regardless of whether it had any intention of owning a satellite, a guaranteed orbital position covering its territory (Canada was allocated three slots). The duration of the Plan is 20 years. The allotted slots are based on rather narrow frequency assignments in the expansion bands (frequencies at C and Ku band that had been added to those available for fixed satellite services at the 1979 WARC).[30] Despite initial hopes that general principles and *a posteriori* rules would be overturned, the outcome was a marginal adjustment which left existing slots and procedures untouched. The concept of multilateral planning conferences to be applied to the whole GSO resource was abandoned.[31] There appears to be some room for the generation of income by owners of GSO slots who are in a favoured location but cannot utilize them themselves, in so far as their leasing to others is not prohibited. Generally, however, there is no direct compensation mechanism within the orbit/spectrum regime. Instead the ITU runs a general technical development programme with UNDP.

There is no central monitoring procedure for adherence to the rules relating to GSO or indeed other frequency use. The IFRB, which is concerned with proper coordination between users, has a tiny staff and will increasingly rely on a computerized system. It is dependent upon notification and reporting by members. This is, then, an essentially self-monitoring system. The obligations on members relate to the proper advanced notification of proposed assignments, which are open to inspection by other members, and the requirement to engage with IFRB in coordination. The Radio Regulations and the Final Acts of ITU conferences are regarded as having the binding force of international law – although the 'footnoting' procedure gives some room for the avoidance of unwanted obligations. There are no formal enforcement or sanctions mechanisms. Adherence to the rules is obtained by 'horizontal enforcement' and by the awareness of users of the mutual dangers of non-compliance in this area. The shared training and values of the engineers who staff national administrations is probably of some importance here. 'The subtle coerciveness of the ITU process lies in the fact that member countries are very interested in being effective partners in the global communications environment (Codding & Landa-Fournais, 1991: 48).

The generation of technical knowledge and its dissemination has always been a significant function of ITU. In terms of the spectrum/orbit regime the pace of technological change has been very rapid and has a central role in allocation, assignment and planning. New methods of frequency compression and reuse have had a significant effect on the carrying capacity of GSO and, it is often argued, can remove the problem of scarcity. The 1985–88 conferences were significantly dependent upon extensive intersessional studies of the feasibility of various planning methodologies and the centrepiece of the 1988 Conference itself was the running of a computer model of the plan. The problem for the ITU system in general, and a stimulus to recent reforms, is that communications are becoming ever more convergent, technological knowledge grows exponentially and 'standard setting' is now a competitive business in which regional bodies and private sector software producers threaten to usurp the organization's established role.

Summary

There is no need to stress the military, political and commercial significance of the space commons. In many ways outer space is directly analogous to the oceans and there are numerous parallels between the development of the law of the sea and the law of outer space. One difference concerns the high cost and difficulty of access, particularly for LDCs, to the valuable and already overcrowded common pool resource of GSO. The impact of the various uses of outer space are universal. Implications for national economic and cultural sovereignty and indeed the political legitimacy of states are becoming ever more evident. Apart from the question of the space environment, the other space-related issue areas and regimes are firmly tied to terrestrial activities and can be seen as the extension of existing military, information flow or frequency spectrum arrangements.

The initial and preponderant uses of outer space were military and demilitarization of the space commons was no more likely than demilitarization of the high seas. Generating some limited common understanding on the deployment and use of anti-satellite weapons was a limited but significant part of the overall attempt to manage the strategic relationship between the United States and Soviet Union – the only space powers. Since the end of the Cold War the military significance of space has been re-emphasized but there has been little incentive to develop multilateral arms control rules within the UN Disarmament Conference machinery.

The condition of the space environment will affect future uses and space debris poses a potentially lethal threat. Here there is evidence of the beginnings of a regime governing orbital practices and spacecraft construction. It is evidently an area of potentially high mutual vulnerability interdependence between spacecraft operators.

The longest running and least successful attempt at regime creation (or change) is provided by the campaign for 'prior consent'. The NWICO campaign, of which space issues were a part, appears to have perished around 1989. The problems and inequities that it addressed have, if anything, become more evident since then, particularly with the arrival of widespread mass-satellite-based broadcasting. The current unregulated regime for communications and remote sensing clearly favours their commercial development by corporations operating on a global scale.

The only area (apart from the Moon Treaty) where there is some evidence of the introduction of common heritage principles is the fundamental one of the management of the orbit/spectrum resource. The regime for GSO, along with the general spectrum regime, is by any standards an impressive achievement in international cooperation. Since the 1970s attempts have been made by the LDC majority in the ITU to redefine the OSR and to introduce the idea of equitable access and national rights to orbital positions. The 1988 agreement to introduce an allotment plan for the expansion bands hardly constitutes regime change, but it does represent a small movement towards common heritage status for the GSO. Equally interesting is the fact it was achieved in the face of initial opposition from developed country users who argued that there was no need for such a plan (because new technologies would overcome any shortage of GSO slots) and that it represented the unwarranted introduction of 'politics' into the running of the regime.

Notes

1. For the details see: Various, 1991, 'Satellites at War: Lessons from the Gulf', *Interavia Space Markets*, 7, 4, pp. 4–13, and Shave, S., 1991, 'CNN Goes to War', *Interavia Space Markets*, 7, 1, pp. 3–7.
2. Owned by a Hong Kong based business consortium involving Hutchison Whampoa and Cable and Wireless, Asiasat is based on a 'second-hand' Westar IV 'C' band communications satellite which was rescued from a failed orbit by the US Space Shuttle. The Hong Kong consortium bought it (refurbished by Hughes Aerospace) at a bargain price from the insurers, and contracted with the Chinese to launch it atop a 'Long March' ELV. This launch occurred in 1990, but only after the necessary clearances had been signed by President Bush in the context of US commercial and political embargoes after the Tienanmen Square massacres of 1989. This was expedited by the fact that Congress was at the time preoccupied with the US intervention in Panama. Asiasat is mainly used for TV distribution but also for communications. Of its 24 transponders, 20 are used to transmit STAR services. Under the terms of the Intelsat agreement, it is not allowed to carry switched international phone traffic. Source: *Intermedia Space Markets* (1990), 6, 3, pp. 139–146.
3. The relevant projects are the US National Aerospace Plane (NASP), the German Sanger and the British HOTOL. They would be true 'spaceplanes' as opposed to reusable launch vehicles like the Shuttle or the ESA's Hermes which relies upon a conventional launcher. Spaceplanes could be flying by the end of the first decade of the twenty-first century.
4. They are Australia, Austria, Chile, Mexico, Morocco and Uruguay. For a legal discussion of the Moon Treaty see Williams (1981). The politics of its rejection is described in Goldman, N.C., 1985, 'The Moon Treaty: Reflections on the Proposed Moon Treaty, Space Law and the Future', in Katz (ed.), pp. 140–149. The discussions of the Treaty in COPUOS for almost a decade mirrored the LoS arguments about the meaning of common heritage. The US view was that it only meant 'equal access' while, for different reasons, the USSR was also sceptical of Third World demands. In the end the USSR introduced a compromise caveat whereby it was understood that 'common heritage' would be interpreted only in the context of the specific Moon Treaty and could not be read in relation to the LoS developments (Goldman, 1985, p. 143).
5. A full treatment of the range of arms control options debated is provided in a report to the US Congress by the Office of Technology Assessment, 'Anti Satellite Weapons, Countermeasures and Arms Control', in OTA (1986).
6. For details see Trux, J., 1991, 'Desert Storm: A Space-Age War', *New Scientist*, 27 July, pp. 30–34.
7. *Facts on File 1993*, p. 66.

8. The Americans also dabbled with nuclear-powered satellites. In 1964 one broke up and released quantities of plutonium into the atmosphere. The details of the failed RORSAT mission are provided in Hart, D., 1987, *The Encyclopedia of Soviet Spacecraft*, Bison/Hamlyn, London, p. 43.

9. Principles involving notification and safety assessment and review drafted by COPUOS legal sub-committee were agreed in 1995.

10. The other way in which spaceflight can affect the terrestrial environment is through the atmospheric effects of the rocket engine exhausts. There has been some concern and evidence of associated ozone depletion, but this was largely discounted in relation to US Shuttle launches by studies related to the Montreal Protocol (Tolba *et al.*, 1992: 431).

11. A small French satellite CERISE collided with a fragment from an Ariane launcher (Wilkinson *et al.*, 1998: 20).

12. Interview, Richard Tremayne Smith, British National Space Centre, January 1999.

13. The estimate of the cost of re-orbiting an Intelsat VI with a 200 km boost is a fuel sacrifice equivalent to one year's station-keeping or $20 million lost revenue. SIG Space Report cited in Chenard (1990: 254).

14. Interview, Richard Tremayne Smith, British National Space Centre, January 1999. It is in this spirit that the Chinese have instituted a debris mitigation programme.

15. COPUOS 'Report of the Scientific and Technical Subcommittee on the work of its thirty-third session (agenda item 5)', GA Supp.20 (A/51/20)1996.

16. For a general survey of the evolution of world communications politics, covering all the main areas of contention from the perspective of their impact upon human rights, see Hamelink (1994).

17. For a detailed history and analysis see Baker (1989).

18. The ITU plan envisaged that DBS would require high power satellites and planned the BSS bands on this basis. However, advances in reception technology meant that it became possible to use lower powered satellites and still obtain adequate reception with a relatively small dish. This enabled FSS frequencies not covered by the plan to be utilized and the 1977 restrictions were thus avoided. British Satellite Broadcasting (BSB) used the nationally allocated frequencies, operated within the plan and operating with a high power satellite was able to use very small 'squarial' antenna. However, by this time Sky had already achieved a decisive commercial advantage.

19. See the *Guardian* 12 March 1994 for the Saudi policy of banning dishes. The other reference is to STAR and Rupert Murdoch, *Financial Times* 14 June 1994.

20. For a discussion of the technicalities of orbital spacing which involve frequency, power levels and bandwidth, as well as antenna characteristics, see Bleazard, G.B., 1985, *Introducing Satellite Communications*, Manchester, NCC Publications, pp. 73–75. The move to higher frequencies allows a separation of three as opposed to two degrees of arc, increasing the notional carrying capacity of GSO to 180.

21. For a classic and near contemporary account see Woolf (1916).

22. Article 3 of the original 1906 Radiotelegraph Convention. Article 8 established an obligation to operate radio services in such a way as to minimize harmful interference. See Jakhu (1983).

23. In terms of GSO this means X band: when the Colombian delegation to the 1985 WARC put out a paper detailing the wastefulness of much of the military occupation of the OSR there was a deafening silence.

24. Specifically the ITU shall 'effect allocation of the radio frequency spectrum and registration of radio frequency assignments in order to avoid harmful interference

between radio stations of different countries' (ITU Malaga Torremolinos Convention, 1973, Art. 4).

25. This was effected at the Extraordinary World Administrative Radio Conference of 1963. In political terms it reflected the very low level of effective participation by the emerging new states of the Third World.

26. A 1961 General Assembly Resolution (1721/XVI) declared that 'communications by means of satellite should be available to the nations of the world as soon as practicable on a global and non-discriminatory basis' (cited in Jakhu, 1983: 399).

27. Res. SPA 2-1, WARC ST, 1971 (see Christol, 1982: 458–459).

28. Article 33 of the 1973 Malaga Torremelinos Convention as amended by the 1982 Nairobi Plenipotentiary Conference reads as follows:

> In using frequency bands for space radio services Members shall bear in mind that radio frequencies and the geostationary satellite orbit are limited natural resources and that they must be used efficiently and economically, in conformity with the provisions of the Radio Regulations, so that countries or groups of countries may have equitable access to both taking into account the special needs of the developing countries and the geographical situation of particular countries.

29. The standard work on the evolution of the ITU system is Codding & Rutkowski (1982).

30. At C band 4.50–4.80 GHz and 6.425–7.075 GHz, i.e. 300 MHz for up and down links, and at Ku band 10.75–10.95 GHz, 11.20–11.45 GHz and 12.75–13.25 GHz, i.e. 500 MHz.

31. The Plan is in Appendix 30 B of Final Acts Adopted by the Second Session of the World Administrative Radio Conference on the Use of the Geostationary Satellite Orbit and the Planning of Space Services Utilizing It (ORB-88) ITU, Geneva, 1988.

6

The Atmosphere

The atmosphere is dynamic and fluctuating. Air masses are in constant movement and the various atmospheric components are continuously subject to chemical transformation and renewal, reacting with solar radiation to provide the essentials of life on the planet. Such 'turnover' is the mainstay of planetary ecology and explains why even small changes in the concentrations of 'trace' gases can have large-scale effects.[1] For example, the concern over 'global warming', which is considered in the latter part of this chapter, arises in the main from an observed change in atmospheric concentrations of carbon dioxide from a pre-industrial level of 280 parts per million by volume (ppmv) to a 1990 figure of 353 ppmv and an estimated rate of annual increase of 1.8 ppmv (Jäger & Ferguson, 1991: 48).

It is not now unusual to describe the atmosphere as a global commons.[2] Nonetheless, the atmosphere does not constitute a commons directly comparable to the high seas, seabed or even outer space. One evident difference might be that a substantial part of it could be regarded as falling within national airspace. As was mentioned in the previous chapter, jurisdiction over airspace extends to the threshold of outer space.[3] Although it has been argued that atmosphere may be part of national sovereignty (Ramakrishna, 1990: 441), this appears legally unconvincing. There is a clear difference between three-dimensional 'airspace' and the 'air' that circulates through it in a way analogous to water moving through nationally owned watercourses (Soroos, 1991: 115). Nonetheless, 'overlap with territorial sovereignty also means that it cannot be treated as an area of common property beyond the jurisdiction of any state, comparable in this sense to the high seas' (Birnie & Boyle, 1992: 390).

There has been at least one (unsuccessful) attempt in the UN General Assembly to describe the atmosphere as part of the common heritage of mankind.[4] Instead the term that is now routinely used in various legal documents relating to the atmosphere and climate change is the 'common concern of mankind'.[5] This expresses the sense that the preservation of the atmosphere is, more than that of any other commons, essential for collective survival.

The global atmosphere is a 'common sink' resource. The preservation of that atmospheric quality upon which life depends must be the ultimate 'public good'. Defining the atmosphere in this way highlights the collective action problem identified by Hume, Olson and many others. There is no simple or proportional relationship between the ownership or control of the sources of atmospheric degradation and the benefits arising from abatement or the costs of inaction. It is impossible to 'enclose' the atmosphere or exclude people from the 'public good' of atmospheric quality. The costs of abatement of atmospheric pollution may have to be borne at one location while most of the benefits may accrue elsewhere.

Changing scientific and public conceptions of the various ways in which the atmosphere is affected by human activities have provided the bases for the construction of atmospheric issue areas. As with the other global commons, most of the problems encountered are a fairly direct consequence of industrialization and the exploitation of technological change. For centuries the atmosphere has been a self-regulating system, maintaining that equilibrium between its chemical components necessary to the preservation of human civilization. Anthropogenic inputs of ever greater amounts of carbon dioxide and other pollutants are now testing the limits of the regulatory capabilities of the system. Awareness of such problems, particularly if impacts are not yet fully observable. is heavily reliant upon scientific understanding, while their formulation as issues occurs within a particular political and economic context.

The atmosphere serves as a medium or carrier of a variety of harmful substances and this provided the focus for early atmospheric issue areas. Articulate public concern with the dangers of radioactive fall-out from the atmospheric testing of nuclear weapons during the 1950s and early 1960s was instrumental in defining an arms control issue area. The outcome was a nuclear testing regime based upon the 1963 'Treaty Banning Nuclear Weapons Tests in the Atmosphere, Outer Space and Under Water'. It was signed by the, then, three nuclear powers, the United States, the Soviet Union and Britain, but was to be disregarded by new nuclear weapons states. A related, but little known, issue area developed around the possibilities of the military uses of environmental modification techniques.[6]

In the 1950s Britain and other countries took legislative action to cope with local atmospheric pollution arising from the unrestricted burning of fossil fuels (smog). It was soon evident that there was an important transboundary dimension to such pollution which required concerted international action at a regional level. Transboundary atmospheric pollution thus became, and remains, a very significant issue area. In fact it would be more correct to speak of a set of regionally based issue areas which have produced transboundary pollution regimes such as that embodied in the 1979 Convention on Long-Range

Transboundary Air Pollution and its various protocols controlling the emission or transboundary fluxes of sulphur, nitrogen oxides and other chemicals.[7] This European regional agreement was, in part, negotiated as a byproduct of improvements in East–West relations under the Helsinki Final Act of 1975.

The conception of global-scale atmospheric problems, involving changes affecting the entire system with potentially universal impacts, has been rather more recent. Two main issue areas have emerged. The first has its origins in scientific speculation as to the destructive impact of chlorofluorocarbons (CFCs), then (in the 1970s) widely used as aerosol propellants, on the stratospheric ozone layer. Increasing evidence as to the extent and causes of ozone layer depletion and its effect on human health led to the relatively rapid creation of an international control regime often named after its most famous legal component, the Montreal Protocol of 1987.

Running parallel to the development of international activity relating to the depletion of the stratospheric ozone layer was the growth of scientific, public and ultimately political concern over the enhanced greenhouse effect and global warming. This was a much more speculative, contested and, in the view of many, ultimately catastrophic matter than any of the other atmospheric issues. There had long been scientific debate about the implications of a build up in greenhouse gases for global warming and climate change. It was only, however, in the late 1980s that this acquired sufficient momentum for political leaders to define a climate change issue area. Substantial diplomatic effort was expended on the creation of an embryo international regime, focused in particular on the control and reduction of carbon dioxide emissions. Its beginnings, as embodied in the United Nations Framework Convention on Climate Change (FCCC), were endorsed by national leaders at the 1992 UNCED in Rio. Since then the development of the regime through the Kyoto Protocol of 1997 has proved to be the centrepiece of international environmental diplomacy.

Stratospheric Ozone

The regime for stratospheric ozone is already seen as a model of international environmental cooperation. The ozone layer, which the regime seeks to preserve and restore, is located in the stratosphere at an altitude of between 15 and 50 km. The stratosphere itself is one of the concentric layers of the earth's atmosphere and lies between the troposphere, nearest the ground, and the ionosphere, which shades away into the vacuum of outer space. The process of ozone production is quite complex. 'Molecular oxygen is broken down in the stratosphere by solar radiation to yield atomic oxygen, which then combines with molecular oxygen to produce ozone' (Tolba & El-Kholy, 1992: 38). In a series of catalytic cycles the ozone is then broken down again. This complicated chemical process has been naturally self-regulating, until the introduction of man-made chemicals (chlorofluorocarbons – CFCs, hydrochlorofluorocarbons – HCFCs, halons, carbon tetrachloride and methyl chloroform) which are now known to deplete the ozone layer. The chemical reactions whereby this occurs were, until very recently, ill-understood. The most significant involves the breakdown of CFCs by the action of UV/B radiation yielding chlorine, which then acts as a catalyst for the destruction of ozone molecules. Other substances such as bromine, a constituent of the halons used in fire extinguishers, are even more destructive but occur in smaller quantities. Many ozone-depleting substances have extremely long atmospheric lifetimes, of up to 100 years.[8]

The consequences of stratospheric ozone depletion are serious because the effectiveness of the latter as a filter for UV/B radiation is diminished. Increases in such radiation, which can penetrate living tissue, entail a variety of dangerous effects. Best known is the increased incidence of skin cancer, but other human health implications include interference with the immune system and increased eye damage, especially through the formation of cataracts. UV/B radiation has also been shown to have deleterious effects, in terms of reduced growth and mutations, for terrestrial plants and aquatic life forms. Thus, as well as posing direct threats to human health, ozone layer depletion also places food production at risk.

The extent and causes of ozone layer depletion were not known with any certainty even at the time of the negotiation of the Montreal Protocol. The formation of a stratospheric ozone issue area and the construction of a control regime has, in fact, run in parallel and interacted with extensive and rapid advances in scientific knowledge. Virtually all of such findings served to increase the gravity and immediacy of the threat. Initially ozone depletion, through the action of chlorine related to CFC emissions, was little more than a scientific hypothesis (the Molina–Sherwood hypothesis of 1974). Invented in 1928 by researchers at Du Pont and General Motors, CFCs were regarded as useful precisely because they were inert substances apparently devoid of harmful side-effects. In the 1970s they had become widely employed as refrigerants, aerosol propellants and for foam blowing and the cleaning of electrical circuit boards. Scientific and public concern, as to damage that CFCs might do to the ozone layer, resulted in 1977 in legislation by the US Congress to ban their use as aerosol propellants. Similar action was taken by Canada, Sweden, Norway, Austria and Switzerland. Together with the United States they were to form the 'Toronto Group' in the subsequent negotiations. From its foundation, UNEP doggedly pursued the ozone problem through international conferences and meetings of experts. Such efforts finally bore fruit with the 1985 Vienna Conference and resulting Convention.

The Vienna Conference and the 'framework' convention definitively established the ozone issue area and committed the mainly developed world participants to a programme of systematic observation and information exchange. It envisaged, but did not contain, control measures. This absence of control measures reflected the division between the two principal groups of protagonists, the Toronto Group and the European Community. There were a relatively small number of participants who, through possessing CFC producing industries, could determine the construction of a control regime. The European Community countries and the United States each accounted for around 36% of global production while Japan accounted for 12%. Furthermore, manufacture was concentrated among a relative handful of major chemical corporations, Du Pont in the United States, Atochem in France and ICI in Britain. Industrial views carried great weight with national delegations, and as Parson (1993: 59) says, the deadlock at Vienna over control measures was entirely predictable. 'Each side proposed international measures that required action on the part of others but not themselves, and indeed in some instances defended their proposals at home by pointing out that they would cost their industries nothing.' The United States, which already had domestic CFC consumption restrictions, wanted their international extension, while the EC favoured a 'production cap'. This would have the effect of preserving the dominant position of EC producers in overseas markets although it would, as EC delegates said, be much easier to enforce than consumption restrictions because of the relatively small number of producers.[9]

Having failed to establish substantive regulations, Vienna did provide a framework committing the parties to further negotiations on a control protocol which would operationalize the aspirations of the Convention. The subsequent negotiations were essentially about finding a compromise that would accommodate the different industrial interests of the key producers. Its achievement, as embodied in the restrictions and phase-out targets of the 1987 Montreal Protocol, owed much to the discovery by Du Pont that there were feasible and indeed profitable substitutes for CFCs.

Neither the impact of CFCs and other substances on the ozone layer, nor the extent of its depletion were in fact firmly established in the formative period of the regime. Public opinion was alerted by the discovery of the Antarctic 'ozone hole', revealed in 1985. The researchers used a balloon to carry instruments that recorded startling levels of springtime ozone depletion. Previous observations by an American Nimbus 7 satellite had apparently been discounted or delayed on the grounds that the findings were so extreme as to be suspect.[10] By the early 1990s it was apparent that severe ozone depletion was not merely a phenomenon observable in the special conditions of Antarctica, but that there was a serious loss above areas of dense human habitation, such as Northern Europe.[11]

The Montreal Protocol regime can best be represented as a dynamic institutional process in which there is a continuous interaction of scientific findings, variable public concern, industrial interest and policy. With UNEP once again taking the lead, the Ozone Trends Panel mounted an intensified study of ozone layer depletion subsequent to the signature of the Protocol. New findings on the damage inflicted by specific chemicals and on the extent and longevity of the problem all continued to press political decision-makers in the direction of tighter regulation and more rapid elimination of the most danger-ous substances.[12] The parties to the Montreal Protocol became engaged in a regularized series of meetings, of which the most important were those held at London in 1990 and Copenhagen in 1992. Two major developments in the regime emerged from this process.

The first involved the extension of participation and the question of North–South compensation. At Montreal there were only 27 signatories and coun-tries such as China and India, both potential major consumers and CFC producers, were notable by their absence. By the 1992 Meeting of the Parties their number had grown to over 100 including India, China and other less developed states. The key decisions concerning LDC participation had been taken in London in 1990, where, over initial American opposition, an Interim Ozone Fund was established to finance the transfer of 'ozone friendly' tech-nology to developing countries which were prepared to abide by the rules of the regime. Without this, the danger would always exist that CFC production would simply be transferred to LDCs not subscribing to the Vienna Conven-tion and Montreal Protocol. The LDC case, as put by Maneka Gandhi, the Indian delegate at London, was unanswerable: 'The West has caused the problem and must help us clean it up'.[13]

The second development involved the extension of the domain of the regime and the tightening of its provisions. The stimulus to such change was con-tinually provided by the scientific community, as more and more evidence pointed to the seriousness of ozone depletion and the involvement of additional chemical substances. It was amplified by public alarm in the developed countries over the extent of the damage and the increased risk of skin cancers. New controlled substances, such as carbon tetrachloride, were brought into the regime. At the same time, often encouraged by unilateral declarations from various parties, the provisions for reduction and phase-out were successively tightened. An indication is provided by the controls for some prominent CFCs. Set at a 50% reduction in 'adjusted production' by 1999 in the Montreal Protocol, a total phase-out by the year 2000 was agreed three years later in London. At the Copenhagen meeting this deadline was moved forward to 1996. This is not to say that the problem has been entirely solved. Yet, although a number of difficulties of implementation and compliance remain and full restoration of the stratospheric ozone layer will only occur at some time after 2050, the regime constitutes probably the outstanding example of contempor-ary international environmental cooperation.

Principles and Norms

The regime related scientific principles to norms of precautionary action and a compensatory approach to the different responsibilities of North and South in what was a quite novel way. Above all there was an awareness of the need to couple regulatory action to an evolving scientific understanding – although this has not always been achieved.

The ozone regime relies upon 'beliefs of fact' – a common and deepening understanding of the relationship between certain chemicals and the harmful depletion of the ozone layer. The preamble to the 1987 Montreal Protocol states simply that 'world-wide emissions of certain substances can significantly deplete and otherwise modify the ozone layer in a manner that is likely to result in adverse effects on human health and environment'. The rather tentative nature of this statement reflected the prevailing uncertainties. Key principles of the regime address such uncertainty. They support an iterative process in which the reduction of scientific uncertainty stimulates and informs policy development, but where the lack of certainty is not to be employed as a restraint on prudent action. This, at least, represents the idealized version of the ozone regime.

The method employed by the framers of the Vienna Convention and Montreal Protocol has frequently been regarded as a paradigm for future global environmental agreements where the scientific bases for action are initially insecure. In one sense this was almost an accident, because the negotiators at Vienna were unable to agree on immediate control measures, opponents citing the absence of conclusive scientific evidence. What emerged was the principle of a 'framework' Convention (Vienna 1985) which establishes the problem and an international commitment to take action in a subsequent and more specific protocol which would be designed to allow frequent updating and amendment. This was to provide an open-ended and adjustable agreement, responsive to changes in scientific understanding of the causes and dimensions of the problem. Benedick (1991: 44–45) has described it as the principle of 'an interim protocol'. Supporting this is an obligation, established in the Vienna Convention (Art. 2a) to cooperate in scientific research and assessment and to share information. Such activities have played a central role in the development of the regime.

Closely allied to the principle of an 'interim protocol' is a behavioural norm of 'precautionary' action. Precaution as a basis for environmental policy was well established in the United States and Germany in the mid-1980s and has already been observed in the development of the marine pollution regime and the Madrid Protocol to the Antarctic Treaty.[14] Weale (1992: 80) has referred to the typical context of precautionary policy, which is exactly that of the evolving ozone regime, where:

> policy-makers are forced to 'go beyond science' in the sense of being required to make decisions where the consequences of alternative policy options are not determinable within a reasonable margin of error and where potentially high costs are involved in taking action.

The essential point is the agreement that action can, and even must, be taken to control a substance even if there is at the time only inconclusive evidence of its damaging effects on the ozone layer. In the case of the Montreal Protocol, subsequent scientific investigation has served to confirm the wisdom of taking precautionary action.

A number of regulatory approaches to the removal or limitation of ozone-depleting substances are conceivable. That adopted, which involves setting timetables for agreed reductions in production/consumption sometimes followed by an outright ban, betrays an implicit principle that the function of the Ozone Protocol is to send signals to the marketplace which will stimulate the necessary innovation to provide alternatives to controlled substances. Thus investment decisions in HCFC technology were prompted by the awareness that under the Protocol CFC production was doomed. Similarly, discussions at the Copenhagen Meeting of the Parties considered suitable dates for the elimination of HCFCs in relation to the lead time for the innovation of alternatives. To put this another way, the corporate sector requires that existing plant and products should be allowed to reach the end of their useful lives, and investments recouped, prior to the phasing out of the offending substances.

The history of the regime also reveals something that may amount to an informal norm in relation to target setting. This emphasizes the desirability of national or regional control activity that goes beyond strict legal requirements. Thus various parties moved ahead of the 1987 Protocol terms by announcing their own more stringent reductions in production and consumption of ozone-depleting substances. The process was repeated at the London and Copenhagen Meetings.[15] That this practice has grown up, does not imply any special altruism on the part of 'advanced' countries. Taking the initiative may not only bring prestige but also commercial advantage.

From the beginning, the necessity for some form of North–South compensation was recognized. Potential producers in the developing world who were not responsible for the problem, but could effectively nullify international restrictions by developing production outside them – had to be accommodated. In terms of the burdens of adjustment to change, the regime espouses the principle of 'equitable control' modified by special treatment for LDCs under Art. 5 of the Protocol, where developed parties are obligated to provide technical and financial assistance. The regime is designed to be inclusive. All are to be encouraged, even induced, to join and there are obligations under the Protocol (Art. 4) to take measures in the control of trade with non-parties. This is obviously essential if the control provisions are not to be circumvented and signatories assured that outsiders will not take advantage of their compliance. The agenda from 1987 to the 1990 Meeting of the Parties was dominated by the question of promoting the membership of prominent developing countries such as China and India. This was finally achieved by arranging a compensatory financial mechanism, the Interim Ozone Fund, which was to be utilized to finance the transfer of non-ozone-depleting technology.

Rules and Procedures

There is no organization dedicated to the ozone regime; rather there are organizational arrangements stemming from the Convention and the Protocol. Decision-making authority rests with the 'Meetings of the Parties' (MoPs). These are state parties to the Vienna Convention (biennial meetings) and Montreal Protocol (annual meetings). Non-governmental organizations (NGOs) and others, including business corporations, may be admitted on an observer basis to Meetings unless one-third of the states present object (Vienna Art. 6.5 and Montreal Art. 11.5). Important Meetings of the Montreal Parties, which have significantly altered and amended regime arrangements, have been held in London (1990), Copenhagen (1992), Vienna (1995) and Montreal (1997). Since the London Meeting there has also been a formal mechanism for the administration of the Interim Ozone Fund, comprising an Executive Committee, composed of an equal number (seven) of developed and LDC parties.[16] There are also a number of significant subsidiary groups reporting to the Meetings of the Parties, including the Bureau of the Conference of the Parties to the Vienna Convention, the Meeting of Ozone Research Managers, the Open Ended Working Group of the Parties to the Montreal Protocol etc. In an important process of institutional development *ad hoc* groups have also been established on compliance, destruction technologies, implementation and data reporting, and there are three expert 'assessment panels' on science, technology, economic questions and environmental effects. A Secretariat provided by the UNEP and based in Nairobi presides over this network of activity.

Rules for collective choice, as expressed in the arrangements for voting and for the entry into force (EIF) of agreements, are quite complex and reflect a number of political disagreements and compromises between the parties. The Vienna Convention established a 'one state, one vote' rule with special provisions for the European Community (a Regional Economic Integration Organization in the Convention's parlance). If consensus could not be achieved, a three-quarters majority of those present and voting, who must comprise two-thirds of the parties, would prevail (Arts 9 and 15). Dissatisfied with the operation of other international agreements operating under such arrangements, where majorities might not include parties with a major stake in the outcome of voting (for example the major CFC producing and consuming countries), the United States pressed for alternative voting arrangements for the Montreal Protocol. This involved 'qualification' of the majorities required on the basis of consumption figures for controlled substances. Thus EIF of the Protocol required at least 11 ratifications from countries consuming two-thirds of estimated global production in 1986. Cancellation of programmed reductions in production and consumption was similarly qualified (two-thirds majority, representing at least two-thirds of consumption). Adjustment of controls towards greater stringency was made easier, requiring only 50% of consumption to be represented in a two-thirds majority.[17]

Distribution rules involving both the extent of reductions and the allocation of burdens are at the heart of the controls in the Montreal Protocol regime. Initially five CFCs were identified along with three halons on the basis of their ozone-depleting potential. The Protocol provided for a 50% reduction in the production and consumption of the CFCs (Annex A Group I substances) by the end of 1999. Using a baseline of 1986, production and consumption was to be frozen by 1990 and reduced by 20% by 1992. The halons (Annex A Group II substances) were to be frozen at 1986 levels by 1992 (Art. 2). Supporting these undertakings were a set of calculation rules that had been subject to painstaking negotiation. In particular, consumption of controlled substances was defined as 'adjusted production': production plus imports minus exports.[18] Some flexibility was introduced into the rules, not only in relation to developing countries, but also to allow industrial rationalization and to permit countries to offset reductions in one area of CFC production and consumption against another. For example, Japan's main interest was in CFC 113, used as a solvent in the electronics industry. The Protocol allowed the continued use of CFC 113 as long as the levels of other CFCs in Annex A Group I were reduced such that the national reduction for the whole group met the interim Montreal targets. All production and consumption of CFCs (for non Article 5 countries) was, under the 1992 adjustment of the Protocol, phased out by 1996. At the same time a range of new ozone-depleting substances has been incorporated and the restrictions on the existing ones tightened. By 1999 no less than 95 ozone-depleting chemicals were 'controlled' under the Protocol.[19]

The control measures apply equally to all parties, except those designated as 'developing countries' under Art. 5 of the Protocol. To be eligible for special treatment such countries must have a per capita consumption figure for a controlled substance of less than 0.3 kg per year. The special treatment involves the right to delay compliance with the Protocol for 10 years after the target dates given for the reduction and phase-out of controlled substances. In association with this, Art. 2 allows parties a 10–15% additional margin on production levels for the purpose of exports to meet the 'basic domestic needs' of Art. 5 countries.

In order to ensure the participation of countries falling under the provisions of Art. 5 a compensatory mechanism was established at the 1990 London Meeting of Parties. A Multilateral Ozone Fund was agreed 'to meet all agreed incremental costs of such parties to enable their compliance' with the control measures. In addition there was a commitment to expedite the transfer of technology under the 'most fair and favourable conditions'. Agreement in London was only made possible by an eleventh hour reversal of US policy that had been hostile to financial compensation. Since the Copenhagen Meeting of the Parties the fund has been operating to pay the 'agreed incremental costs' incurred by developing countries in phasing out their consumption and production of ozone-depleting substances. It had, by 1999, disbursed some $903 million.[20]

The designers of the regime were well aware of the 'public goods' type characteristics of a restored ozone layer and the consequent 'free rider' problem. Commitment to the control provisions might well erode if there was any suspicion that other parties were taking commercial advantage by avoiding their implementation. The monitoring of compliance is, therefore, a central pillar of the ozone arrangements. The control measures were themselves designed in such a way as to rest on (supposedly) easily obtainable and verifiable statistics. Early on in the negotiations the EC representatives pointed out the impossibility of monitoring consumption, by thousands of users, as opposed to production (Benedick, 1991: 79–92). Thus the concept of consumption in the Protocol relies upon production adjusted by imports and exports – both apparently easy to establish.

Parties to the Montreal Protocol are subject to fairly strict injunctions in terms of adherence to agreed control measures and obligations to report data on compliance.[21] Unlike the provisions of the spectrum/orbit regime, but in common with other contemporary environmental agreements, reservations are specifically not allowed. However, there is a continuing problem with the successive amendments to the regime which are legal instruments requiring the slow process of ratification.[22]

Article 7 of the Protocol places an obligation upon parties to report data on annual production, import and export of controlled substances along with data for the baseline year 1986. After some initial problems with the fulfilment of this obligation an effective data reporting and implementation review system has developed (Greene, 1993a, 1993b, 1998). 'Up to 1996 . . . compilations of national data have indicated good overall progress in implementing commitments – mostly in advance of Protocol obligations – and have revealed no significant compliance problems except with reporting obligations themselves' (Greene, 1998: 94). Problems have, however, been encountered with the data reporting of the successor states of the USSR and the developing country parties.

In common with most of the regimes that have been considered, compliance cannot be compelled and there is a reliance on 'horizontal enforcement' stemming from a concern for a loss of reputation in the eyes of other Parties. There is also the possibility of utilizing the compensatory funding mechanisms. In the mid to late 1990s there were problems centring on the acknowledged non-compliance of Belarus, Russia and Ukraine. These were handled by the regime's evolving non-compliance procedure, centring upon the Implementation Committee. A satisfactory outcome was achieved by involving the Global Environmental Facility of the World Bank. Victor (1998: 164) concludes: 'While the Committee applies mainly the soft-management approach to non-compliance, it has been effective in its most difficult cases of non-compliance only because it has access to the slightly "harder" tools of conditionality'. With regard to non-parties there are definite enforcement provisions within the Protocol. Under Art. 4 exports and imports of controlled substances are banned and within three years of EIF trade in articles containing controlled substances as well.

The principles that underpin the regime ensure that the generation of knowledge and information has been fundamental. Articles 3, 4 and 5 of the Vienna Convention enjoin the parties to cooperate in relevant research and to transmit information. Under the Protocol Art. 9, this is extended to the promotion of public awareness of the stratospheric ozone problem. The interaction between the various scientific assessment panels organized by the Secretariat and the Meetings of the Parties is well established and provides one side of the central dialogue between science and policy that is supposed to drive the regime.

Viewed over the first decade of its existence, the most impressive feature of the regime has been the way in which its development has responded continually to new evidence on the depletion of the stratospheric ozone layer and the role of particular chemicals in the process. The figures for actual reductions in consumption are notable. In 1986 consumption of CFCs was approximately 1.1 million tonnes world-wide. By 1997 this had been reduced to 146 000 tonnes. Consumption by developing countries has increased over the same period, but in the most significant users has started to decline. There remain a range of problems to be tackled: the rise in atmospheric concentrations of halons (used in fire extinguishers), widespread smuggling of CFCs, and the export of CFC-based refrigerators to developing countries which threatens to create a continuing illegal market for this primary ozone-depleting substance. However, the success of the regime would be difficult to dispute. One critical note concerns the pace of the regime's construction. Although progress has been vary rapid since 1985, it took eight years, during which the ozone layer was rapidly depleted, to move from an expression of international concern in 1977 to the agreement of the Vienna Convention, which added very little of substance to that contained in the previous document (Parson, 1993: 72).

Climate Change

The 'greenhouse', or more properly 'enhanced greenhouse' effect and the arguments about the probability of global warming and associated climate change are now very well known. Greenhouse gases (ghgs) include carbon dioxide, methane, nitrous oxide and the halocarbons covered under the Montreal Protocol. (Also covered under the Kyoto Protocol are three industrial gases: HFCs, PFCs and SF_6.) Although they are only trace elements in the atmosphere they have the important property, unlike oxygen and nitrogen, of interfering with the passage of energy. Thus, they have always acted to absorb heat radiated from the earth's surface – the greenhouse effect. Historically the result has been the maintenance of a stable mean global surface temperature, an essential condition of the development of the ecology of the planet as we know it. The problem arises from the enhanced greenhouse effect. Human activities, and notably the generation of excessive amounts of carbon dioxide through the combustion of fossil fuels, have increased and will continue to increase the atmospheric concentration of ghgs with a variety of predicted effects in terms of temperature rise, climatic change and sea level rise. These, in turn, are associated with a whole range of possible 'impacts' including the inundation of coastal regions, desertification and the radical alteration of global patterns of land use, food supplies and even disease.[23]

Climate change as an issue differs in a number of very significant ways from that of stratospheric ozone depletion, even though some of the same gases are involved. There is the question of immediacy. Ozone depletion had clearly occurred and represented a threat to the existing human population, while climate change through the enhanced greenhouse effect is still disputed and will only be fully observed well into the next century. If the scientific predictions are to be believed, concerning the kind of environmental disasters that may occur in a world inhabited by our great-grandchildren, then novel issues of intergenerational equity arise.

'May' is of course the operative word, for another key characteristic of the climate change issue is scientific dispute and high levels of uncertainty. While the depletion of the stratospheric ozone layer could actually be monitored and was very soon not in serious doubt, climate change remains an hypothesis (despite some evidence of warming in recent decades) heavily reliant on simulations of the future. The reliability of the methods used by climate modellers, and the extent to which their findings should be taken seriously, have been matters of hotly contested debate which have on occasion divided not only technical experts but governments. This serves to emphasize the desirability of obtaining a source of unprejudiced scientific advice representing something like a consensus amongst specialists. The Intergovernmental Panel on Climate Change (IPCC) may represent the closest approach to such an impossible ideal but, as will be discussed below, the idea of a completely depoliticized and disinterested science is a chimera.

Another essential difference concerns the origin and complexity of the problem. In retrospect, ozone depletion had simple causes and readily available solutions. With climate change, it is not a matter of controlling a handful of specialized chemicals produced by a limited number of companies for which (in the main) acceptable substitutes could be found. The most essential and universal of human activities give rise to the release of over 6 billion tonnes of carbon dioxide per annum into the atmosphere (Hulme *et al.*, 1992: 5). Methane and nitrous oxide, while naturally occurring, also have anthropogenic sources in agriculture and the burning of biomass and fossil fuels.[24] In terms of the contribution of these different emissions to the increased 'radiative forcing' which is responsible for temperature changes, the IPCC estimates are that carbon dioxide accounts for 55% of the problem, while methane and nitrous oxide account for 15% and 6% respectively (Houghton *et al.*, 1990: xx, fig.7). There is still a large measure of uncertainty regarding the precise role of the various greenhouse gases, especially in relation to sinks, which include the oceans and forests. The latter have been highlighted both as stores of biodiversity and as a vital part of the carbon balance – acting as they do to fix and photosynthesize carbon dioxide.

Additional complexity arises from the differential distribution of responsibility for ghg emissions on the one hand and probable impacts on the other. We may speak of a global problem, but this, as was argued in Chapter 1, should not imply some form of common and proportional responsibility or indeed a risk that is equally shared. In terms of carbon dioxide emissions there is a gross asymmetry between the contributions of developed and less developed economies. Measured in per capita tonnes for 1988, each US citizen's share of carbon dioxide emissions is 5.3, each British, 2.7, and each Polish, 3.8. In contrast the Chinese and Indian citizen is responsible for 0.6 and 0.2 respectively (Tolba & El-Kholy, 1992: 68).

Neither is there a clear sense that all are at equal risk from potential climate change. It was possible to assert a rough equality of risk in the case of stratospheric ozone depletion, but the impacts of global warming are even more difficult to calculate than its likely extent and there may even be benefits for, to provide just one example, the colder regions of Russia which at present do not support agriculture.

All this is not to say that the problem is insoluble, only that it is likely to be very difficult. According to the authors of the sequel to the controversial *Limits to Growth* computer simulation of the early 1970s, a pattern of 'overshoot and collapse', partly involving the effects of global warming but also driven by the demands of a rising population, is not inevitable. Avoidance requires a 'comprehensive revision of policies and practices that perpetuate growth in material consumption and in population . . . and a rapid drastic increase in which materials and energy are used' (Meadows *et al.*, 1992: xvi). The problems are awe inspiring, not least in the sense that it is difficult to abstract the atmospheric commons, either in analytical or policy terms, from a huge range of other issues and human activities ranging from demographic change to the functioning of the international monetary system.

As Skolnikoff (1993: 183) has observed, 'global climate change is the apotheosis of the idea that "everything is related to everything else"'. While recognizing this, it is still the case that a necessary, but hardly sufficient, part of any solution is to be found in international cooperation directed towards a common reduction of sources in terms of ghg emissions and preservation and even extension of sinks. Unless this occurs national remedial action will be dogged by the kind of collective action problem so familiar in the other global commons.

The attempt to create a regime relating to climate change, and more specifically the regulation of ghg emissions, is in its infancy. Its origins can be traced to the way in which UNEP, and to a lesser extent WMO, gave political salience to a growing but far from certain scientific concern with global warming. The idea of global warming is an old one but international attention, amongst climate modellers at least, can be traced back to the 1970s. In 1976 WMO and ICSU launched a Global Atmospheric Research Programme and in 1979 the First World Climate Conference was convened in Geneva. A consensus as to the role of the various ghgs began to emerge by the time of a scientific meeting in Villach, Austria, in 1985. Yet it is important to stress that this was still a matter for specialists, an issue area in political and policy terms had yet to be defined. This seems to have finally occurred during 1988 when, for a variety of reasons, 'global warming' seized both public and political attention. The consequence was a flurry of high-profile meetings involving political leaders as well as scientists. The issue was even afforded a prominent place in the communiqué of the 1989 G7 Paris Summit.[25]

As with the development of the Montreal Protocol, with which the FCCC is intertwined in a number of ways, the basis of political action was necessarily some degree of international scientific consensus on the dimensions of the problem. UNEP and WMO sponsored this through the setting up in 1988 of the Intergovernmental Panel on Climate Change which reported to the 1990 Second World Climate Conference. The IPCC spoke with authority because the scientific conclusions of its Working Group I (if not the other Working Groups) could be represented as the collective opinion of 'most of the active scientists working in the field' world-wide. Their prediction was that under a 'business as usual scenario' in which no action was taken to curb anthropogenic emissions of ghgs, there would be a mean global temperature rise of 1°C by 2025 and 3°C by the end of the twenty-first century. The corresponding predicted rises in sea levels were 20 cm by 2030 and 65 cm by the end of the century (Houghton et al., 1990: xi). This proved to be enough to overcome the reluctance of a number of developed countries to take action and the initiation of negotiations towards the conclusion of a framework convention was agreed by the UN General Assembly in December 1990. The negotiations, involving over 100 states, were conducted by the Intergovernmental Negotiating Committee (INC) under the auspices of the General Assembly and utilizing its secretariat rather than UNEP. Five sessions were held from February 1991 through to the conclusion of a draft convention in May 1992, ready for signature as planned at the UN Conference on Environment and Development held at Rio in the subsequent month.

Negotiations were far from easy and the final text of the FCCC is laced with ambiguities reflecting the need to accommodate the often wide disparities and disagreements between participants. Amongst the industrialized countries national positions inevitably derived from calculations as to energy efficiency and the costs of reducing ghg emissions.[26] US policy reflected the high level of national carbon dioxide emissions and the perceived costs to an economy in recession of attempting any immediate reductions. There was an initial tendency to discount scientific estimates of the scale of the climate change problem, which was subsequently replaced by the advocacy of a 'comprehensive' approach to sources and sinks that would counterbalance the stress placed by other negotiators on carbon dioxide and implicit US responsibility. The European Community painstakingly edged towards a common policy of the stabilization of carbon dioxide emissions at 1990 levels by the year 2000, although it failed to agree a common approach to implementation through a Community system of carbon taxation. A commitment to the target agreed by the Community became a central component of the INC negotiating draft but proved absolutely unacceptable to the Bush administration in an election year. Last-minute diplomacy was required to persuade the Americans to sign a revised version of the FCCC which, to widespread dismay, omitted this commitment.

Aside from the disagreements between G7 countries, North–South issues figured strongly in the negotiations. They involved the demand for recognition of the legitimacy of development priorities in the context of Northern responsibility for the climate change problem; the question of the scale of compensatory finance if the South was to adhere to any agreement; and the matter of the relationship between the monitoring and enforcement of any agreement and national economic sovereignty. Further complexity was added by the specific interests of a range of coalitions ranging from the oil-producing states, naturally wary of any restrictions on the burning of hydrocarbons, through to the Alliance of Small Island States, whose very existence was threatened if the IPCC's predictions concerning the implications of 'business as usual' for ghg emissions were accurate.

The 1992 Framework Convention on Climate Change (FCCC), which finally emerged from the INC process, represented only the beginnings of a climate change regime. It formally entered into force on 21 March 1994 and the first Conference of the Parties (CoP) met in Berlin in spring 1995. This meeting set itself the task of devising control measures before the end of 1997 – the 'Berlin Mandate'. After two years of *ad hoc* negotiations a Protocol to the FCCC was agreed at the 1997 CoP at Kyoto.

Principles and Norms

A recognition of the likelihood of anthropogenically produced climate change is fundamental to the regime. It sets the objective of the Convention which is to ensure 'stabilisation of greenhouse gas concentrations in the atmosphere at a level that would prevent dangerous anthropogenic interference with the climate system' (Art. 2). The FCCC is explicitly a 'framework' convention, following the precedent set by the LRTAP and ozone regimes, wherein development of control measures is informed by the advance of scientific understanding and periodic review of the 'adequacy of commitments'. The 'precautionary principle' is to be applied so that a 'lack of full scientific certainty' should not be used to impede the adoption of necessary measures (Art. 3.3). Although initially focused upon the control of ghg emissions, it is clearly stated that 'sinks' as well as sources are to be taken into consideration, although the extent to which one might be offset against the other has remained controversial.

The underlying allocative principle of the regime is expressed in terms of 'common but differentiated responsibilities and capabilities'. This reflects an understanding, expressed in the preamble, that the problem originates with the developed countries. For the purposes of the Convention they are defined in Annexes I and II. Annex I includes 36 industrialized countries, including those like Russia which are 'undergoing transition to a market economy', but excluding the Asian non-industrialized countries and Latin America. Annex II is more selective and contains the members of the OECD, the European Union, the United States, Japan, Canada, Australia, New Zealand, Iceland and Turkey.[27] Developing countries, currently making a low contribution to global warming in per capita terms, will inevitably increase their emissions in the future, but action by them is expressly made contingent upon the fulfilment of the commitments of the developed countries (Art. 4.7). This principle was to become increasingly contentious as the regime developed. The argument was advanced that Southern economies would not only be responsible for an ever-growing proportion of global emissions but were already competing with and undercutting developed world manufacturing industries.[28]

All Parties, including developing countries, are enjoined by Art. 2.1 to perform a range of tasks involving the compilation of data and the development of policies to mitigate climate change. However, the developed Parties of Annex I have more immediate injunctions directed to them, consistent with the principle that they should 'take the lead in combating climate change and the adverse effects thereof' (Art. 3.1). It had been a widespread hope, prior to the 1992 UNCED, that these would include a binding obligation to reduce emissions of ghgs to 1990 levels by the year 2000. Yet because of US objections, Art. 4.2a of the Convention merely urged the developed Parties to adopt such policies and measures that would demonstrate that:

developed countries are taking the lead in modifying longer-term trends in anthropogenic emissions consistent with the objective of the Convention, recognizing that the return by the end of the present decade to earlier levels of anthropogenic emissions of carbon dioxide and other greenhouse gases not controlled by the Montreal Protocol would contribute to such a modification.

There is a related commitment to provide details of their national policies and measures and data on their sources and sinks (Art. 4.2b). The other main obligations of the developed Parties involve the provision of financial assistance and technology transfer to non Annex I countries (Arts 4.3, 4.5) and particular assistance to the most vulnerable countries in coping with the adverse impacts of climate change (Art. 4.4).

In meeting their commitments to emissions reductions the developed countries are allowed a degree of 'flexibility'. As will be discussed below, the mechanisms outlined in the Kyoto Protocol opened the door to various investment and trading opportunities whereby developed countries might find alternative ways of meeting their ghg reduction commitments. They may be regarded as the embodiment of another regime principle, the promotion of a 'supportive and open international economic system that would lead to sustainable economic growth and development in all parties' (Art. 3.5).

Decision-making

The supreme decision-making body of the Convention is the Conference of the Parties (CoP), which has met annually since 1995. It will also serve as the 'Meeting of the Parties' to the Kyoto Protocol (Cop/MoP). CoPs are now very substantial gatherings attracting hundreds of official participants and thousands of lobbyists and NGO activists. CoP4 (1998 Buenos Aires) was, for example, attended by over 5000 people including representatives of 170 governments.[29] One peculiarity of the CoP is that it continues to operate without formally agreed rules of procedure. Such rules were on the agenda at CoP1 in Berlin but a refusal by oil-exporting countries to countenance majority voting and by other Parties to hand an effective veto to the latter led to an impasse that has persisted. Business has instead been conducted under draft rules, which, with the exception of voting rules, are simply 'applied'. The absence of voting places the onus upon the Chair to decide whether a matter is agreed or not (Grubb & Anderson, 1995: 1). A pattern has emerged whereby sessions (in working groups or the Committee of the Whole or COW) where national officials haggle over disagreed text within square brackets are succeeded by a 'high level segment' of the Conference at which ministers and even Vice Presidents apppear. Inevitably, critical decisions are delayed until the final moment and there is, beforehand, a good deal of posturing and 'hostage taking' to provide material for final concessions. If the complaints from some participants are to be believed, most Parties are excluded from the negotiations that matter. These occur in private informal sessions of contact groups involving the the major developed countries, the EU and representatives of the G77/China (*Earth Negotiations Bulletin*, 1998: 27).

Reporting to the CoP are two subsidiary bodies, for implementation (SBI) and for scientific and technical advice (SBSTA). In order to assist the development of the regime, two temporary bodies were created by CoP1 at Berlin in 1995. The Ad Hoc Group on the Berlin Mandate (AGBM) met regularly to negotiate what was to become the Kyoto Protocol and the Ad Hoc Group on Article 13 (AG13) met to consider the unresolved problem of ensuring implementation. Article 13 of the FCCC 'Resolution of questions regarding implementation' actually contains only one sentence which requires the Parties to consider the establishment of a multilateral consultation process. The whole apparatus is serviced by an FCCC Secretariat which was set up in Bonn.

Rules

Since the fulfilment of the Berlin Mandate by the Kyoto Protocol, agreed at CoP3 at the end of 1997, the outline of the rules of the climate regime has emerged with much greater clarity. Standards of behaviour for all Parties are listed in Art. 2 of the Protocol which prescribes 'policies and measures'. Included are measures to promote energy efficiency, enhance sinks, promote sustainable agriculture, remove market imperfections relating to fossil fuels and reduce emissions from bunkers. The crux of the Kyoto Protocol is, however, in its rules for the distribution of the burdens of reducing ghgs. This responsibility, arising from the original Art. 4.2 commitments of the FCCC, remains exclusively a matter for Annex I countries, despite attempts to introduce 'voluntary commitments' by others.[30] Quantified emissions limitation and reduction objectives or QUELROs, as they became known, had been argued over since the 1980s and having failed to include any binding objectives in the FCCC their provision was the primary purpose of the 1995 Berlin Mandate agreed at CoP1 and argued over in AGBM meetings over the next two years. Before Kyoto the EU proposed a 15% reduction from a 1990 baseline by 2010, while Japan proposed 5% with the United States appearing to offer little more than a standstill. Also under active discussion in the AGBM and at Kyoto was the question of which gases to include (either a three or six gas basket) and, most controversially, the extent to which there could be 'flexibility' in achieving any targets that might be set.

The compromise agreed at Kyoto included both QUELROs and flexibility mechanisms, although the operational rules for the latter were left for subsequent CoPs to develop. The collective goal of the Annex I countries is a global reduction of approximately 5% in relation to 1990 baselines, in the 'first commitment period of 2008–2012. This is to be achieved by a net reduction of ghgs arising from cuts in emissions and removals by sinks (Kyoto Art. 3).[31] A differential allocation of responsibilities for achieving the Kyoto target is set out in Annex B to the Protocol. Under this 'big bubble' the majority of the developed countries, including the EU, have to achieve 8% reductions; the United States 7% and Canada 6%. Some countries, such as Australia and Hungary, are allowed actual increases (8% and 10%) while Russia and the Ukraine are allowed to stand still.[32]

Agreement would have been impossible without the provision of what became known as the Kyoto 'mechanisms', which allow flexibility in the achievement of national commitments. These consist of joint implementation (Kyoto Art. 6), the clean development mechanism (CDM) (Kyoto Art. 12) and, most controversial of all, emissions trading (Kyoto Art. 17). Joint implementation allows the transfer and acquisition of emissions reduction units between Annex I parties in the context of joint projects to reduce emissions or enhance sinks, but on the understanding that any credited reductions must be additional to those that would otherwise occur and supplemental to domestic action (Kyoto Art. 6.1).[33] An alternative means of acquiring emissions credits is provided by the CDM. Under Art. 12 of Kyoto, Annex I countries can obtain them through participation in projects with developing countries, based upon 'certified emission reductions accruing from such project activities'.[34]

The third Kyoto mechanism is emissions trading. There have been various theoretical attempts to elaborate a global system for the management of ghgs. These have ranged from a common carbon tax to a sophisticated market-based tradeable permits system.[35] Article 17 of the Kyoto Protocol represents the first formal expression of international intent to institute such a system. Developed countries may engage in trading in order to fulfil their Art. 3 commitments; such trading shall be 'supplemental to domestic actions'. In common with the other mechanisms, the all-important 'principles, modalities, rules and guidelines' are left to subsequent CoPs. Of particular importance here is the question of 'additionality', the extent to which emissions reductions achieved by the mechanisms really will be additional to, as opposed to a substitute for, domestic action. There is also a related requirement to avoid transactions in 'hot air' – emissions credits which arise from prediction or accounting errors and which do not represent actual reductions in ghgs.[36]

None of the emissions reduction obligations agreed at Kyoto are undertaken by developing (non Annex I) countries, China and the G77 having successfully resisted attempts to make further progress in the climate convention dependent upon the adoption of such commitments, whether 'voluntary' or not. Nonetheless, non Annex I countries do have reporting and other obligations under the FCCC[37] and there was from the beginning a compensatory aspect to the regime whereby it is understood that 'new and additional' financial resources will be provided to 'meet the full agreed costs' incurred by developing countries in carrying out their commitments (Art. 4.3). There is also an explicit statement that the extent to which developing countries will effectively implement their commitments will depend upon the fulfilment of developed country commitments in terms of technology transfer and financing (Art. 4.7). The latter will be provided through the Global Environmental Facility (GEF) of the World Bank. This was not uncontroversial given the misgivings of the G77 and the attacks by environmental activists on the Bank's previous disregard for the environmental consequences of its lending. Thus, it took until 1998 to agree that the GEF, on the basis of a restructuring of its activities and work programme, would be the financial mechanism of the Convention.[38]

The commitments to emission reductions at Kyoto relate to the period 2008–2012 and the supporting mechanisms will not be operative until 2001 at the earliest. In the meantime, from the entry into force of the Convention, it has been functioning as an information generation and review system. All parties are obligated under Art. 4.1 and Art. 12 to provide national communications on their policies and measures to mitigate climate change and inventories of anthropogenic emissions by sources and removals by sinks. There are, however, different timescales for the provision of national information. Developed countries are required to communicate within six months of Entry into Force, others within three years and the least developed are allowed to report at their discretion. It is one of the primary functions of the CoPs to review such national communications – utilizing the subsiduary bodies. The scope of the problem, in terms of devising agreed methodologies for the measurement of emissions from sources and the absorptive capacity of sinks, is probably unprecedented in the history of international cooperation. The very nature of the Convention and considerable methodological uncertainiies would make formal non-compliance procedures and penalties inappropriate. Instead, the process of implementation review represents a form of 'horizontal enforcement' based upon transparency and peer pressure. To institutionalize this a multilateral consultation procedure has been developed under Art. 13 of the FCCC with an accent on the provision of advice and assistance.[39]

The situation with the Kyoto Protocol is very different. Here, it is recognized that a robust non-compliance procedure is required involving an 'indicative list of consequences taking into account the cause, type, degree and frequency of non-compliance' (Kyoto Art. 18). Accurate measurement, verification and enforcement of the rules is central to the operation and, indeed, acceptability of the 'mechanisms'. There is already strong apprehension that if they are loosely drawn and monitored they will amount to little more than a means whereby developed countries can circumvent their domestic emission reduction obligations. Very considerable sums of money may be at stake in future emissions trading allied to a considerable potential for fraud involving inadequate measurement and accounting. The adequacy of verification appears as, or probably more, important here than any enforcement penalties on those who wilfully break the rules to be established for trading or joint implementation. Throughout the Convention there is a recognition that its development will be shaped by 'research and systematic observation' (Art. 5). The specific mechanism for adjusting the regime to the development of scientific understanding of climate change is periodic review of the 'adequacy of commitments' (Art. 4.2 (a, b and d)). The first such review took place at CoP1 in 1995 and provided the basis for the Berlin Mandate. A second review was required before November 1998, but CoP4 failed to arrive at any conclusions on how the Convention might be further extended in the light of increasing scientific certainty as to the anthropogenic causes of climate change. The debate was inevitably caught up with the question, which dominated the Buenos Aires meeting, of 'voluntary commitments' by developing countries (*Earth Negotiations Bulletin*, 1998, 12, 97: 16). Existing commitments and the measures set out in the Kyoto Protocol are unlikely to impose the necessary restraints upon the rise of ghg emissions in the long run. The CoP will have to revisit the 'adequacy of commitments' considering not only the extension of Annex I country commitments but the thorny problem of developed country commitments relating to what can only be an ever-increasing share of global emissions.

Achieving a politically credible understanding of what is necessary in the highly contentious field of climate change research relies upon establishing scientific estimates, which, if not based upon full consensus, have sufficient weight and authority. The Intergovernmental Panel on Climate Change (under the auspices of the WMO and UNEP) was established in 1988 for this express purpose. Although it has an 'arms length' relationship to the FCCC, its findings being considered *inter alia* and subject to the interpretation of the SBSTA, it must be considered as an integral part of the regime. The SBSTA is composed of multidisciplinary but government-appointed personnel and is charged with the assessment of scientific evidence and the provision of policy advice (although not the identification of innovative technologies). Its policy role sets it apart from and serves, to some extent, as an insulator for the IPCC which is precluded from making policy recommendations.

The 'intergovernmental' title is not accidental as the panel is composed of government-nominated experts. They are organized into three working groups: WGI, The science of climate change; WGII, Impacts, adaptation and mitigation; and WGIII, Economic and social dimensions. The end product of this process of scientific evaluation which probably includes the majority of those actively involved in climate change research on a global scale is a series of Assessment Reports. The first was published in 1990, the second in 1996 and the third is scheduled for 2001. Their remit is to provide a consensus account of the state of knowledge on climate change. It is this which provides both their authority and, on occasion, the kind of opacity associated with a carefully negotiated compromise text. As Brack & Grubb (1996: 1) note, the IPCC has 'evolved into what is probably the most extensive and carefully constructed intergovernmental advisory process ever established in international relations'.

The 1990 first Assessment Report is 'formally recognised as the transition point between exploratory discussions on how to deal with greenhouse warming and formal negotiations towards an international convention' (Nitze, 1990: 2). During the INC negotiations the IPCC reported to the negotiating sessions on such matters as the reassessment of the contribution of CFCs to global warming. The 1996 Second Assessment Report confirmed that human activities are changing the atmospheric concentrations and distributions of ghgs (Houghton *et al.*, 1996).

IPCC draws from a bewildering range of international research efforts such as the World Climate Research Programme and the International Geosphere–Biosphere Programme. Some idea of the complexity of the 'alphabet soup' of international scientific cooperation can be drawn from Figure 6.1. The diagram illustrates how bodies (such as SCAR) which we have already encountered in the other global commons regimes are interconnected in the enterprise of researching what is often termed 'global change'.

The development of IPCC and the international scientific network to which it is connected is in itself a fascinating case study of the interaction between science and policy, which, as in other areas, is never simply a matter of one-way traffic where political decision-makers receive the objective wisdom of their scientific counterparts (Boehmer-Christiansen, 1993, 1996). The IPCC Chairman hinted as much in his report to the Second World Climate Conference where he spoke of the need not to 'mix up' the scientific assessment tasks of the IPCC with the inevitable political considerations that would figure in the negotiations for a convention. 'It is obvious that a clear division

---

Figure 6.1 *The organization of international scientific research into global environmental change. (Source:* Initiatives, Programmes and Organisations, *UK Research Councils Global Environment Research Office, volume 3, 1994). ATS = Antarctic Treaty System; GCOS = Global Climate Observing System; GEF = Global Environment Facility; GEMS = Global Environment Monitoring System; GOOS = Global Ocean Observing System; ICSU = International Council of Scientific Unions; IGBP = International Geosphere–Biosphere Programme; IHDP = International Human Dimensions Programme; IHP = International Hydrological Programme; IOC = Intergovernmental Oceanographic Commission; IPCC = Intergovernmental Panel on Climate Change; ISSC = International Social Science Council; MAB = Man and Biosphere Programme; SCAR = Scientific Committee on Antarctic Research; SCOPE = Scientific Committee on Problems of the Environment; SCOR = Scientific Committee on Oceanic Research; UNDP = UN Development Programme; UNEP = UN Environment Programme; UNESCO = UN Educational, Scientific, and Cultural Organization; WCP = World Climate Programme; WCRP = World Climate Research Programme; WHO = World Health Organization; WMO = World Meteorological Organization*

145

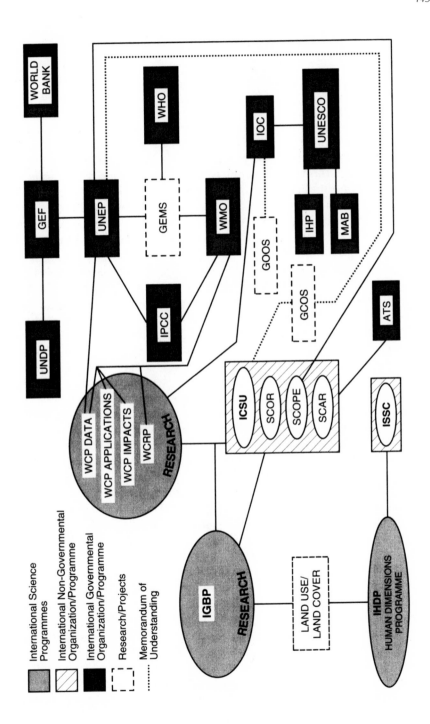

of responsibilities in this way was not quite achieved in the final stage of the IPCC assessment' (Bolin in Jäger & Ferguson, 1991: 20).

Summary

Partial and local regimes relating to atmospheric issues have existed since the 1960s, but it was only during the 1980s that there was both a realization of the extent of global atmospheric problems and a serious political attempt to come to grips with them. The function of global atmospheric regimes is not to distribute some form of common international property but rather to provide a framework which will encourage and enable governments and corporations to behave responsibly in curbing their use of the atmosphere as a common waste disposal system or, as in the case of the ozone regime, to phase out the use of substances that are found to do particular damage.

The regime for the protection of the stratospheric ozone layer has achieved paradigmatic status as an example of good regime building practice. The word 'building' is used advisedly because this was an institution self-consciously created from first principles over a relatively short period. The contrast may be drawn with the accretion and codification, over hundreds of years, of norms and rules for the use of the oceans, or over decades in the case of the utilization of the frequency spectrum. Much of the urgency derived from quite sudden scientific discoveries and shocks relating to the extent of the ozone 'hole' and the need to take immediate action if the stratospheric ozone layer was to be restored. The method, pioneered by the ozone regime, of an initial framework convention followed by an interactive process of scientific discovery and the enactment and extension of controls has been widely recommended.

It is, therefore, not surprising that the attempt to lay the foundations of a climate change regime should reflect recent experience with the Montreal Protocol. There are a number of other connections between the two regimes, including an overlap of their domains. Under the ozone regime CFCs are being replaced by substitute HCFCs (both ghgs). Yet in the Kyoto Protocol, developed country Parties are committed to the emissions reductions of HCFCs.

The similarities are greatly overshadowed by the sheer scale and difficulty of the problem of coping with the enhanced greenhouse effect and the many fundamental ways in which it differs from what seem, in retrospect, the simplicities of the ozone issue. For, as Skolnikoff (1993: 193) notes, 'the extent of difficulties encountered in banning CFCs when so few interests, relatively, were at stake does not offer much encouragement'. It is difficult not to strike an apocalyptic note when discussing climate change and present regime building efforts are meagre indeed. For pressing political reasons matters such as the fate of forests or demographic change do not form part of the climate change issue area as presently constructed. Carbon dioxide emissions and the five other ghgs have been the focus of attention and even here the measures in the FCCC and its Kyoto Protocol can at best be described as being close to the starting point on a long but urgent journey. Probably the most decisive determinant of progress will be the most elusive: the level of popular interest that is aroused. Judging by the ozone experience, effective public pressure on governments will probably only derive from clear and incontrovertible evidence that global warming is actually occurring. The pessimistic conclusion is that, according to some authorities, by that time environmental changes may already be irreversible.

Notes

1. A total of 99% of the earth's atmosphere is composed of two gases: nitrogen (78%) and oxygen (21%). The remaining 1% includes argon and such 'rare' gases as hydrogen, carbon dioxide, methane and oxides of sulphur and nitrogen (there are also liquids and particles in suspension).
2. Pearce *et al.* (1991), for example, includes a chapter (Ch. 2, pp. 11–30) entitled 'The Global Commons', which is mainly devoted to economic approaches to global warming with a brief section on the conservation of biodiversity. Caldwell (1990) also regards the atmosphere as a global commons. There is a more recent discussion in Soroos (1997: 213–229).
3. As discussed in the previous chapter, there is no precise and agreed legal boundary between outer space and sovereign airspace, but for practical purposes the limit may be set at the lowest possible satellite orbit – strictly speaking a perigee with an altitude of around 100 miles (160 km).
4. By Malta in the debate leading to Resolution 43/54, cited in Birnie & Boyle (1992: 391).
5. See, for example, the Ministerial Declaration of the Second World Climate Conference, 7 November 1990 (Jäger & Ferguson, 1991: 536–539), and the FCCC itself which begins with the phrase: 'Acknowledging that change in the Earth's climate and its adverse effects are a common concern of mankind'.
6. In 1976 a 'Convention on prohibition on military or any other hostile use of environmental modification techniques' was signed by 55 nations. There are also the UNEP 'Provisions for Co-operation Between States in Weather Modification' of 1980, which attempt to cope with the problem of the transboundary effects of non-military experiments in weather modification.
7. The relevant agreements are the Helsinki Protocol of 1985 on sulphur, the Sofia Protocol of 1988 on nitrogen oxides, and the Geneva Protocol of 1991 on volatile organic compounds.
8. The various substances are ranked in terms of their ozone-depleting potential in Annex A of the Montreal Protocol. CFC 11 has an ozone-depleting potential of 1 while Halon 1301 has a potential of 10.
9. The issues, from a US point of view, are well summarized by that country's chief delegate, Benedick (1991: 77–97). His book also provides a graphic first-hand account of the negotiations. A critical corrective from the point of view of Benedick's British counterpart is provided in a review article by F. McConnell in *International Environmental Affairs*, 3, 4, Fall 1991, pp. 318–320.
10. See Parsons (1993: 31).

11. The European Stratospheric Ozone Experiment reported a loss of ozone of up to 20% over Northern Europe at the beginning of 1992 (*Guardian*, 8 April 1992).
12. There are good accounts of the scientific background, which form the basis of the brief discussion in the text, in Tolba & El-Kholy (1992: 32–59) and Thomas (1992: 199–237).
13. *Financial Times*, 28 June 1990. For a discussion of the development issues see Thomas (1992: 228–235).
14. Benedick (1991: 24) claims that the precautionary principle shaped both US policy and the general approach of the Vienna Convention and Montreal Protocol. He further claims that it was unfamiliar at the time to most European officials and 'perplexed EC negotiators'. This seems strange because the 'Vorsorgeprinzip' was well established in German practice and informed the creation of the EC's Fourth Environmental Action Programme. See Weale (1992: 79–84). Regardless of the question raised by Benedick, this seems a good example of the way in which domestic policy norms are transferred to the international level.
15. Montreal Protocol Art. 2.11 establishes a right to take measures beyond the terms of the Protocol.
16. See Annex IV, Appendix II, of the London Revisions 1990. Voting is by a two-thirds majority which must comprise a simple majority of both the developed and LDC group.
17. Benedick (1991: 88–91) covers the US position on voting and the negotiations on the Montreal Protocol. The EIF provision is in Montreal Protocol Art. 16. Article 2 covers the procedure for the cancellation of second phase 30% reduction in CFCs, which must involve a two-thirds majority representing two-thirds of 1986 estimated consumption. The threshold for adjustments towards greater stringency was set at a two-thirds majority representing half of global consumption. This was subsequently amended at London to a majority of the parties operating under Art. 5 (LDCs) and a majority of the remainder (London Revisions H. Art. 2) The above are 'adjustments' and would become binding once having received the required majority. On the other hand the addition of new substances would constitute an amendment governed by the voting requirements of Art. 9 of the Vienna Convention, requiring formal ratification by two-thirds of the parties (Montreal Protocol Art. 2.10).
18. The controversy over calculation of control levels set the European Community, the largest CFC exporter, against the United States. The EC initially wanted a production cap which would tend to 'lock in' its dominance of export markets. As Benedick (1991: 79–92) recounts, the eventual calculation arrangements in Art. 3 were a compromise.

    The control levels refer to aggregate figures for groups of substances listed in Annexes to the Protocol. The calculated level is also dependent upon the ozone-depleting potential of each substance within a group given in relation to CFC 11 which has a potential of 1.0, CFC 113 thus is deemed to have less impact with a potential figure of 0.8, while Halon 1301 is rated at 10.
19. The revisions to date are:

London 1990

Annex A
5 Group I CFCs – Complete phase-out by 2000.
3 Group II halons – Complete phase-out by 2000; plus 50% reduction by 1995.

New Annex B – Using a 1989 baseline

Group I
10 new CFCs – Complete phase-out by 2000; plus 20% reduction by 1993 and 85% reduction by 1997.

Group II
Carbon tetrachloride – Complete phase-out by 2000; 20% reduction by 1995.

Group III
Methyl-chloroform – Complete phase-out by 2005. Frozen by end 1993; 30% and 70% reductions by 1995 and 2000.

New Annex C
'Transitional' HCFCs noted but not limited.

Copenhagen 1992

Annex A
Group I CFCs – Complete phase-out by 1996; 75% reduction by 1994.

Annex C
HCFCs – 'Significant reductions' by 2020; complete phase-out by 2030.

Annex E
Methyl bromide – Developed countries to freeze production at 1991 levels by 1995.

Vienna 1995

Annex E
Methyl bromide – Phase-out by 2020.

Montreal 1997
Finalized the schedules for the phase-out of methyl bromide.

20. London Revisions 1990. Art. 10 and 10a. An interim fund worth $240 million was created in 1990 to cover the period until the EIF of the London Revisions (10 August 1992). At the 1992 Copenhagen Meeting of the Parties the Multilateral Ozone fund was established with funding of $340–500 million for 1994–96. The United Kingdom and other developed countries proposed that the fund should be incorporated within the World Bank's Global Environmental Facility – a suggestion stoutly opposed by Southern parties.
21. Montreal Protocol Art. 2 on control measures and Art. 7 on data reporting.
22. In mid-1999, there were 168 Parties to the 1987 Montreal Protocol, 129 to the 1990 London Amendment, 88 to the 1992 Copenhagen Amendment and 11 to the 1997 Montreal Amendment.
23. There is now a very extensive literature on global warming. Useful accounts of the scientific background are to be found in Leggett (1990), Tolba & El-Kholy (1992: Ch. 3) and Thomas (1992: Ch. 5, authored by Matthew Paterson).
24. Each of the main ghgs have different radiative forcing characteristics, expressed as global warming potential (GWP). For a 20 year time horizon, and assigning carbon dioxide a GWP of 1, methane is 63, nitrous oxide is 270 and CFC 11 is 4500. However, carbon dioxide and methane are still seen as the main problems

because despite their lower GWP they occur in the greatest volume. Source IPCC findings in Jäger & Ferguson (1991: 33).

25. For a good description of the emergence of climate change as an issue see Paterson (1996a: 16–48) and also Brenton (1994: 163–195).

26. There is a tabular summary of national positions in Thomas (1992: 182–183). On the issues under negotiation and the various positions see also Moss (1991: 16–29).

27. The economies in transition included in Annex I are allowed a 'certain degree of flexibility' in meeting their commitments (Art. 4.6). They are also not required to participate in the Annex II countries' obligation to provide assistance to developing countries under Arts 4.3–4.5.

28. The principle of 'common but differentiated responsibilities' was taken to mean that for the purposes of fulfiling the Berlin mandate at least, developing countries would not be required to undertake any new commitments. However, at the behest of the Senate, which passed the Byrd–Hagel Resolution in 1997, the US government made participation in the Kyoto Protocol dependent upon the assumption of commitments by developing countries. The Resolution (which passed 95-0) made it plain that the US should not be a signatory to any protocol that excluded developing countries from binding commitments or that harmed the US economy. Such commitments were rejected by the G77/China at both Kyoto and CoP4 held in 1998 at Buenos Aires. At the latter meeting Argentina was encouraged to offer a voluntary commitment – a move followed immediately by US signature of the Kyoto Protocol (*Earth Negotiations Bulletin*, 1998, 12, 97: 29).

29. There were 70 ministers and 1500 national officials in attendance, plus some 2600 observers from IGOs and NGOs, and also some 880 members of the press (UNEP Information Unit for Conventions, Press Release 14 November 1998).

30. Notably at CoP4 Buenos Aires 1998 where US signature of the Kyoto Protocol was synchronized with Argentinian adoption of voluntary commitments, amidst strong objections from the rest of the G77.

31. One point of difficulty here is the inclusion of sinks relating to land use change and forestry (LUCF).

32. The 'bubble' concept was promoted by the European Union in its approach to the negotiations and endeavours to accommodate the different levels of development of countries within the achievement of a common ghg reductions target. High-energy users contribute greater reductions than those who are less industrialized and who may be allowed actual increases from a low baseline. It is recognized by Kyoto Art. 4 which sets out rules for Parties agreeing to jointly fulfil their commitments. If there is a failure to achieve the combined target of the bubble then each Party will be responsible for achieving its nationally assigned level of emissions. While all EU members have national emissions targets of an 8% reduction in ghgs, the actual reductions (or increases) that they will make were agreed by the Council of Ministers in June 1998 and differ widely: Austria −13%, Belgium −7.5%, Denmark −21%, Finland 0, France 0, Germany −21%, Greece +25%, Ireland +13%, Italy +6.5, Luxemburg −28%, Netherlands −6%, Portugal +27%, Spain +15%, Sweden +4%, UK −12.5%. The aggregate effect will be to provide an overall reduction of 8% for the EU as a whole. Thus, the EU's position in relation to the Kyoto Protocol might be described as a 'bubble within a bubble'.

33. Joint implementation projects had already been investigated in a pilot projects phase preceding Kyoto, reviewed by the SBSTA. The 1998 CoP4 decided to continue such 'activities implemented jointly under the pilot phase' (6/CP.4).

34. Note here the 'banking clause' which allows emissions reductions achieved before

2008 to counted. Some developing countries are in favour of a CDM bank because they would have more control over it than the GEF.

35. For example see Nitze (1990) on a carbon tax and the work of Grubb (1992) for the most extensive exposition of the tradeable permits idea.

36. CoP4 held at Buenos Aires in November 1998 did little to elaborate the mechanisms but did set out an extensive work programme of institutional, methodological and process issues that were to be resolved through discussion in the SBSTA and SBI, prior to CoP6 in 2000 (7.CP4, FCCC/CP/1998/16/Add.1).

37. Article 4.1 sets out the commitments of all Parties. Alongside the preparation of national inventories and policies, Parties are also charged with cooperation in technology transfer, sustainable management, scientific research, information exchange and education. They are also to cooperate in addressing the impacts of climate change and are to take the latter into account 'to the extent feasible' in all relevant policy areas.

38. The GEF was mentioned as the interim financial mechanism of the Convention in Art. 21 of the FCCC, and it was finally established by Decisions 2 and 3 of CoP4. The financial provisions only apply to developed countries listed in Annex II, i.e. those which are not 'in transition to a market economy'. The financial mechanism is dealt with in Art. 11, but only in general terms. The phrase 'new and additional' is significant because of arguments surrounding the Montreal Protocol Interim Fund where the United States had argued that 'new' could mean funds transferred from other items in existing aid budgets.

39. The procedure was developed by the Ad Hoc Group on Article 13 and adopted by CoP4 at Buenos Aires in 1998.

7

Regime Effectiveness

The question of effectiveness should be at the heart of any discussion of regimes. Even to employ regime approaches implies that the analyst has taken up a position in terms of the debate, discussed in Chapter 2, on whether regimes matter. The assumption that guides the present work is that they do matter, but that there are significant differences across the various global commons issue areas. At first sight some issue areas are 'governed' by apparently well developed and successful regimes, while in others identifiable institutions are absent or weakly developed. A key explanatory task is to account for these variations. Such an exercise is likely to have, whether explicitly or implicitly, a normative purpose. This will often be the improvement of existing arrangements such as to avoid 'commons tragedies' or promote 'efficient' management. Yet such intentions are hardly unproblematic or uncontested and may neglect the important equity criteria embodied in notions of 'common heritage'. In much of the literature there is also a wholly understandable tendency to concentrate on successful examples of good practice and to explore the ways in which existing institutions may be 'improved'.[1] This provides one good reason for including weak regimes, failures and 'non-regime' situations in a comparative analysis. Indeed, it is difficult to discuss the incidence and effectiveness of regimes at all without the reference points provided by cases where they are absent or ineffective.

The argument that there is a need to consider the incidence of regimes and explain relative success and failure still assumes, that there is:

> some mystery about the rather uneven performance in recent times of many international arrangements and organizations. While some lie becalmed and inactive, like sailing ships in the doldrums, others hum with activity, are given new tasks, and are recognized as playing a vital role in the functioning of the system. (Strange, 1983: 341)

This 'mixed record' is explained by Strange in terms of the three different purposes of regimes. They can be strategic – the instruments of the strategy of a dominant state, adaptive or symbolic. While some of the organizations and arrangements under consideration here have 'symbolic' elements, they are overwhelmingly 'adaptive' in character, that is to say they arise from the new problems associated with the growth of the world economy and the advance of technology which have 'also often enlarged the possibility of reaching agreement as well as the perceived need to find a solution' (Strange, 1983: 342).[2] It may be argued that, for our purposes, the important variations exist within this category of adaptive institutions and that there is no simple relationship between the existence of a commons problem and the emergence of a regime that will provide a 'solution'. What criteria should be used to compare the performance of 'adaptive' regimes for the commons?[3] Because the detail of formal regime arrangements has long been the professional preserve of the international lawyer, it is not surprising that the standard measure of effectiveness has often been legal. The focus has been upon the principles, norms and rules themselves, upon what might be called the internal characteristics of the regime. This would comprise such matters as the extent to which the rules are binding and have treaty status, participation and provisions for entry into force, arbitration of disputes and future adaptation and codification. Of particular interest are the mechanisms that are provided to ensure compliance with the treaty obligations undertaken by states. There is a substantial and growing technical literature on these matters.[4]

Students of international relations and organization have been more concerned with the closely related issue of how far, if at all, regime arrangements transfer authority from a national to an international level. In their view, the key to the provision of effective institutions does not reside with the specific characteristics of the rules or legal instruments as such, but with the creation of what Leonard Woolf (1916) and many others optimistically called 'international government'. The belief that this is not only desirable but possible, through the consent of the international community, has inspired 'idealist' or 'rationalist' and 'functionalist' thinkers. A contemporary example of policy advocacy in this tradition is provided by the Report of the Commission on Global Governance (1995: 251–253). The very distinguished and experienced members of the Commission call for a trusteeship of the global commons 'exercised by a body acting on behalf of all nations'. This is to be achieved by transforming the old Trusteeship Council of the UN, which has largely exhausted its original responsibilities in terms of 'trust' and dependent territories. It would evidently centralize responsibility for the management of the commons but there is no discussion of any additional powers that might be granted to the Council.[5] In the light of our initial discussion of the problem of the governance of the commons there are compelling reasons for assessing regime effectiveness in terms of the extent to which rules and decisions are internationalized. Thus alongside and complementing essentially legal criteria, the orthodox International Relations approach might be dubbed 'effectiveness as the transfer of authority'. Its focus would be on the national or international character of norms and rules and above all upon the way in which 'the making of collective choices' is organized. If regimes are a form of governance somewhere along a continuum stretching between unbridled national independence and an hypothetical world government, the measure of effectiveness would be the distance travelled towards the latter.

Both the above approaches would be incomplete without reference to the question of the effect of institutions in modifying behaviour at the national or corporate level. For the participants in a major transnational research project on the implementation of international environmental commitments, effectiveness is to be measured by 'the extent to which the accord causes changes in the behavior of targets that further the goals of the accord' (Victor *et al.*, 1998: 6). In this sense the technical legal compliance of states with their international obligations may only be a means to an end and it is quite likely that situations will arise where compliance is high but effectiveness is low. The extent to which regimes do modify individual and collective behaviour is, as we have already seen, a central point of contention in the debate between the various paradigms. In much of the 'rationalist' or institutionalist writing there is often a rather easy assumption that if the correct rules and organization can be erected, the intentions of the regime will be translated into reality. On the other hand, the neglect of institutional factors in favour of an exclusively power-based analysis, characteristic of realist commentaries, may be equally misplaced.

We may, thus, describe a third dimension of effectiveness in terms of 'behaviour modification' attributable to regimes. Until the 1990s this evidently significant question received surprisingly little attention in the literature (Haggard & Simmons, 1987). This shortcoming has now been extensively addressed in a number of studies (Haas *et al.*, 1993; Underdal, 1994; Bernauer, 1995; Victor *et al.*, 1998). Their focus is the causal relationship between institutions and behaviour, the ways in which regimes determine or fail to determine national or lower level decisions. This involves not only questions of monitoring and enforcement, which have always been essential to discussions of collective action problems in the management of commons, but also the more subtle and non-coercive ways in which the existence of institutions may condition and even alter the perceptions of national and corporate actors. There is also a practical recognition that most of the governments in the international system are deficient in the means to implement their international commitments. For this reason 'capacity building' through the transfer of resources, technology and expertise has a very significant place in recent environmental treaty making.

The foregoing approaches to effectiveness concentrate on regime rules and decision-making procedures. None of them, either separately or collectively, can provide a wholly satisfactory set of effectiveness criteria. Common sense dictates that ultimate effectiveness must relate to the problem under consideration and the extent to which it may be 'solved' by a regime. This must involve the relevance of the underlying purposes of the regime in terms of norms and principles. It can be the case that the most sophisticated and developed legal and organizational architecture, which may under strictly formal criteria be deemed effective, superintends a 'non-problem'. The principles of the regime may be misplaced or it may be erected on the basis of an 'issue area' which is far too narrow. Making such judgements will involve the observer with a set of performance criteria that are external to the regime itself.

Ultimately the most significant of such effectiveness criteria will be those relating to the fate of the commons and questions of their sustainability and equitable management. However, these are also the most difficult criteria to establish because opinion is both divided and shifting. All that can be done is to attempt to establish some independent reference points, reflective of whatever scientific consensus may exist, in the full knowledge that they are hardly likely to be definitive. 'Effectiveness as problem solving' will thus be based on a judgement of what is required for the maintenance and development of the commons and a corresponding assessment of the performance of the various regimes. In the end this will lead back to the modification of behaviour, for such is the role of institutions.

This chapter will discuss these varying conceptions of effectiveness in turn. The fundamental purpose will be to arrive at effectiveness criteria that can be used to rank the various commons regimes under consideration. It also provides an opportunity to draw the material in the previous four chapters together in a comparative analysis.

Effectiveness as International Law

Legal scholarship highlights the significance of the internal architecture of the regime, the way in which its rules are constructed and the consistency with which they relate to norms and general principles of international law. The international law of the environment and the commons is viewed as undergoing a continuous process of development and refinement. At the level of norms and principles this would entail, for example, the elaboration of the 'precautionary principle'. Lawyers will also discuss in quite narrow and technical terms the types of rule or procedure that are most likely to prove efficacious in commons management and represent the best current practice. The effectiveness of rules is measured in terms of their legal status and quality and the extent to which they are considered binding upon sovereign states. In this respect 'compliance' has a restricted meaning denoting the extent to which the behaviour of state parties conforms to explicit treaty rules (not to principles and norms) (Mitchell, 1996: 5). Furthermore, modern environmental agreements will often contain a compliance system involving procedures for the multilateral review of state adherence to rules (Werksman, 1996: 85–121). Beyond this, there is a growing realization that much more needs to be known about the extent to which international and domestic legal systems interact. The modification of the latter by the former may represent a critically important function of international regimes.

Perhaps the most authoritative, but at the same time constricted, legal view of effectiveness is that of the UNCED Prepcom. During 1991 it established an open-ended Working Group (no. III) to deal with legal and institutional questions. It devised no less than 32 separate 'criteria for evaluating the effectiveness of existing agreements or instruments'. These were almost exclusively legal in character and hardly ever strayed into questions touching the political determinants of effectiveness or the real environmental impact of agreements. Effectiveness criteria involved, *inter alia*, the extent of formal participation, implementation, information, reporting, review and codification. The subsequent report, which dealt in detail with the extent to which 124 existing agreements met the criteria, provides the best extant example of a comprehensive legal approach to effectiveness and covered most of the regimes surveyed in this book (Sand, 1992).

Formal participation by states in the various regimes was a key issue for the UNCED survey, with an understandable emphasis upon the factors promoting the membership and effective participation of developing countries. While this reflects the UNCED agenda, universal participation may not be a measure of effectiveness for global commons regimes. In all cases other than the restricted Antarctic Treaty System, membership is open to all states and legal participation will occur when a state ratifies one of the formal global commons agreements. Yet, as may be observed from Table 7.1, for most of the agreements under consideration such membership is far from universal. Effective participation, in the sense of playing a full role within the various regimes rather than being a 'mere signatory', is in fact much more limited than the ratification figures would suggest. Participation poses a range of complex difficulties for developing countries in fulfilling data collection and reporting requirements and in serious involvement in technical discussions. Adequate numbers of trained and experienced staff may simply not be available. The ITU may enjoy near universal membership but its involvement in negotiations on issues which are only comprehensible to radio engineers provides an extreme case of such difficulties.[6] Both in the development of the stratospheric ozone regime and the nascent climate regime, some attention has been paid to such problems and the need to provide funding, technical assistance and special terms of entry for LDC participants.

A further and very significant trend in the last 20 years has been the great increase in NGO participation and the movement in most regimes from

Table 7.1 International legal agreements for global commons: selected aspects

	Full members	% Developing states	Ratifications for EIF	Signature to EIF – months	Amendment requirements	Reservations allowed?
ICRW – Whaling 1946	40	4	6	23	Three-quarters	Yes
Antarctic Treaty 1959	43	39	All	19	All	Yes
Outer Space Treaty 1967	96	78	5	8	Simple majority	Yes
London Convention 1972	77	58	15	21	Two-thirds	Yes
MARPOL Convention 1973 and 1978	106	69	15*	119	Two-thirds	Yes
ITU Nairobi Convention 1982	188	80	55	13	Two-thirds	Yes
UNCLOS III 1982	133	77	60	143	Two-thirds or 60; three-quarters for Seabed	No
Vienna Convention – Ozone 1985	169	78	20	44	Three-quarters	No
Montreal Protocol – Ozone 1987	168	77	11**	15	20	No
FCCC – Climate 1992	173	78	50	21	Three-quarters	No
Kyoto Protocol – Climate 1997	8	100	55***	–	Three-quarters	No

Notes:

* Plus 50% tonnage

** Plus Annex I Parties emitting 55% of total Annex I 1990 emissions

*** Plus two-thirds consumption CFCs

Sources: Sand (1992) and IWC, AT, IMO, OOSA, ITU and UNEP/IUC Websites. Membership figures are for mid-1999. The definition of developing/developed countries is derived from Annex 1 of the FCCC.

exclusion to the provision of observer status.[7] This is most dramatically illustrated by the Antarctic Treaty Consultative Party meetings which remained for a long time as closed and exclusive gatherings of state representatives but have recently expanded to include NGOs.[8]

Arguably, regimes operate more efficiently with a restricted membership composed exclusively of those who have real interests in and control over the relevant issues (this view is further developed in Chapter 8 below in relation to structural approaches to regime creation and change). One finding of the UNCED survey was that developing countries were in fact underrepresented in agreements dealing with 'operational' as opposed to 'declaratory' aspects of environment and sustainable development (Sand, 1992: 11). The virtue of universal participation, well understood in the preparations for UNCED, is that it serves to legitimize the regime. Anything other than the encouragement of formal participation by all states would not, nowadays, be practical politics within the UN system. The difficulties encountered by the exclusive Antarctic regime, created before the great expansion of UN membership in the 1960s, are instructive in this regard.

Lawyers contemplating regime effectiveness will understandably be concerned with the status of norms and rules and their internal coherence. At the highest level this will involve the legal validity of regime norms and principles. For example, whether the common heritage of mankind constitutes *jus cogens*, a norm recognized by the international community of states from which no derogation is permitted, is a matter of continuing legal dispute. Similarly there is no definitive acceptance of the concept of permanent state sovereignty over natural resources (Schrijver, 1997: 221–222, 374–377). By contrast, the doctrine of 'reasonable use' of common property would be generally acceptable (Birnie & Boyle, 1992: 121).

The coherence of regimes, in terms of the way in which specific rules follow from well-grounded principles and norms, provides another criterion of legal effectiveness. Here, there is some variation amongst the cases. At one extreme the Antarctic regime has developed incrementally and does not display an orderly progression from first principles to detailed regulation. At the other, a more recent piece of law-making, as embodied in the ozone regime, has a consistent legal architecture and is seen to be much more satisfactory.

Perhaps the most important legal concern is with the status of rules. It is, after all, the authoritative character of particular rules and customs that give them the status of law. A central question in legal discussions of effectiveness is how to provide binding rules which are also flexible enough to cope with rapid change, particularly in the field of environmental regulation. The most binding type of regime rule is embodied in the form of treaty law, the solemn undertakings of sovereign states. Most of the regimes under consideration do have basic legal instruments of this type (the Antarctic Treaty, the Vienna Convention or the International Convention on the Regulation of Whaling).

Where such legal instruments do not exist, as in the case of the mitigation of orbital debris, there is usually an ambition to create them. However, the effectiveness of treaty law may be vitiated by the time and trouble associated with the ratification procedures that are required prior to entry into force (EIF). The various ratification targets are given in Table 7.1 and vary widely. Experience with the 60 ratifications required for EIF of the Third Law of the Sea (LoS) Convention, highlights the problem of achieving a balance between a ratification target that is small enough to allow early EIF, but not so limited as to allow a regime to come into being without the consent of potentially important participants. The ozone regime, which along with the FCCC has relatively high-speed EIF provisions, met this problem by including Art. 16 of the Montreal Protocol which requires 11 ratifications by nations accounting for two-thirds of global consumption. Similarly the MARPOL has an EIF clause that requires at least 15 ratifications constituting at least half of world gross merchant tonnage. The Antarctic regime, however, requires unanimous approval of treaty amendments and for the EIF of protocols. These provisions might be said to have been effective in a peculiarly inverse way, for the painstaking negotiation of a Minerals Convention concluded in 1988, came to nothing when two of the Consultative Parties, Australia and France, refused to accede. The Madrid Protocol (1991), which amongst other things imposes a 50 year ban on mining, took until 1998 to enter into force although the Parties had already agreed to abide by its provisions in the interim.

The legal provisions of recently created regimes also have the virtue of disallowing reservations. By taking out a reservation (or using an objection procedure) a state can sign and ratify an agreement while indicating that it will not be bound by a particular part of it. Perhaps this facility, which is a feature of older established treaty systems such as that of the ITU, the ICRW and the London Convention where it has been used and the MARPOL where it has not, was necessary when legal instruments required unanimous or near unanimous ratification. One reason advanced for the prolonged failure of the Third LoS Convention to achieve EIF was the fact that states were not allowed under its terms to protect their specific interests through reservations. The existence of a reservations procedure (used no less than 118 times at the 1989 Plenipotentiary) has not, however, meant that the ratification of ITU Conventions has always been prompt.[9]

Although Treaty Law is still regarded as the most effective form of rule, the difficulties outlined above and the way in which regimes now have to respond to a rapidly changing knowledge base, has meant that international environmental lawyers have been much concerned with so-called 'soft law' alternatives. These involve 'either moving more slowly towards the formalization of obligations or of setting goals for conduct that, although informal, are intended to have some authoritative status' (Birnie & Boyle, 1992: 30). Interestingly, this tends to align with the regime concept as understood in International

Relations, where norms and rules can take a variety of forms and exist beyond the 'black letter' law of international treaties. Of course, customary law always existed and provided the basis for uncodified but authoritative regimes, such as that which 'governed' the high seas for centuries. An example of the utility of 'soft law' is provided by the Recommendations of the Antarctic Treaty Parties which, while lacking the binding quality of treaty law, appear to have been widely, if not universally, observed.

The placing of operational rules within an annex to a Treaty or Protocol, where the requirements for revision and amendment are not so stringent as for the main legal instrument itself, is now recognized as playing a significant role in the promotion of regime effectiveness. The 'state of the art' in terms of current treaty-making practice, to ensure speed and flexibility of regulation in response to evolving scientific advice, is the 'framework convention – adjustable protocol' model pioneered in the development of the ozone regime and adopted in the FCCC.

A real test of flexibility relates to a function of all the global commons regimes, in the generation and incorporation of knowledge. The ozone regime has been held up as a paradigm case of rapid accommodation to evolving scientific knowledge. In its early years the whaling regime, with its failure either to generate the necessary scientific understanding or to respond with regulations to such clear indications of the collapse of cetacean populations as were available, represents the other extreme. Although it has always been heavily concerned with technical knowledge, there must also be doubts about the efficacy of ITU structures in this regard. Chayes & Chayes (1991: 282) point out that the ITU as a 'first generation' international organization is hampered by its need to acquire the express consent of the parties to changes.

Some form of collective decision-making procedure is a defining characteristic of a regime. In formal terms this will usually involve the regular Conferences of the Parties or equivalents such as ITU Plenipotentiaries and Radiocommunication Conferences. They operate within a legal framework of 'standing orders' and voting rules which necessarily reflect the fundamental principle of international law – sovereign statehood. The fullest expression of absolute state sovereignty would be to give every participant an effective veto by requiring consensus as in the ATS model. Much of the UN system has, however, adopted a one state one vote simple majority system (as in ITU) which, because of the difficulty of binding a sovereign state against its will, is tempered by either giving any decision the status of a recommendation or by permitting reservations. In the FCCC Conference of the Parties there has been a continuing dispute over draft rules where some Parties have objected to the possibility of being overruled by majority voting. This has meant that during the 1990s the Conference of the Parties operated without formally adopted rules of procedure. Although 'one state one vote' majority systems may have a rather spurious form of democratic legitimacy (equating the vote of India with that of Andorra), the legal form bears little relationship to underlying economic and power structures. For this reason, various weighted systems (as for example in IMF or Intelsat, where votes are related to contributions) have been developed. The best example amongst the various commons regimes is provided by the complex arrangements for voting by the various 'chambers' in the Seabed Authority.

In actual practice the difficulties involved with any form of voting have led to a strong predisposition towards the achievement of consensus – reflected in the wording of many of the treaties under consideration where it is explicitly stated that voting should be regarded as a last resort. Voting systems are not as vestigial as the legal provisions for dispute settlement, which although spelled out in extensive detail in the various treaties, appear to have been rarely utilized (Sand, 1992: 14). As will be argued in Chapter 8, voting systems, even if not utilized, can play a significant role in regime politics. Nonetheless, it is not plausible to argue that any particular voting system will, in itself, tend to make a regime more or less effective in terms of its responsiveness to change. Formal procedures are merely one factor amongst a number which will bear upon the making of collective choices.

The foregoing brief discussion only begins to consider the legal dimensions of global commons agreements. Monitoring and compliance, considered in a subsequent section of this chapter, provide another area of considerable legal interest, as does the under-researched but vital topic of the relationship between regimes and the modification of domestic law.

The UNCED survey correctly places a great deal of emphasis on 'transparency' and information sharing as a measure of regime success. Yet, in the words of one reviewer this also involves it in 'the wholly inadequate suggestion . . . informed by a legalistic perspective . . . that an agreement is effective merely if it is reviewed regularly'.[10] It is easy to share such irritation with the narrowness of much legal discussion and even to arrive at the conclusion that such technicalities can be disregarded in any serious discussion of regime performance. Such a view would reflect one side of the reciprocal neglect that tends to characterize the relationship between the disciplines of International Law and International Relations or Political Economy. Just as apolitical discussions of legal frameworks will tend towards unreality, so the significance of legal questions for the study of regimes should not be underestimated. Any close account of the political processes of regime formation (as for example provided by Benedick, 1991) will provide ample evidence that legal questions are taken extremely seriously by governments. Many of those closely involved in negotiations are lawyers by training and key 'sticking points' in the formulation of agreements often have a legal character. The precise terms of agreement remain significant not only because they can incur costs for the acceding state but also because they provide the basis upon which others can be held to account.

Effectiveness as Transfer of Authority

While the idea that regimes constitute a form of governance at some point along a continuum, bounded at one end by interstate 'anarchy' and at the other by a hypothetical supranational world authority, is widely shared, there have been few attempts to explore exactly what the stages on the continuum might be. One such attempt has been made by Donnelly (1986) in a study of human rights regimes. He provides a simple typology for assessing regime 'strength', which is measured in terms of the extent to which authority over an issue area is transferred from a national to an international level. This transfer can occur on two dimensions: norms (Donnelly makes no distinction between principles, norms and rules) and decision-making procedures, which provide the axes of the diagram in Figure 7.1. His typology has a number of potential uses for the comparative study of various regimes and indeed their development over time.

Four types of norm are identified, running from 'fully international to entirely national'. 'National standards: the absence of substantive international norms' equate to the absence of any rule other than the pursuit of self-interest and would represent the situation, frequently encountered in earlier discussions, in which there is an unregulated commons. The first rung on the ladder towards full international regulation is represented as 'international guidelines: international standards which are not binding but nonetheless widely commended'. This would encompass a variety of forms of 'soft law', Antarctic Treaty Recommendations and UNEP environmental guidelines. The standards and recommended practices (SARPs) under development for the mitigation of orbital debris also reside in this category. The next category is one that covers 'international standards with self-selected exemptions'. ITU Regulations for frequency and orbit fall into this category, because reservations and 'footnotes' are specifically allowed, as do the ICRW rules with their loopholes and objection procedures. The highest category is represented by 'authoritative international norms: binding international standards generally accepted as such by states'. Arguably very few, if any, of the regimes under consideration would completely meet such exacting requirements. Yet we may argue that the principles and norms relating to the preservation of the stratospheric ozone layer and, to a lesser extent, the need to mitigate the effects of climate change might be included. Historically, the Law of the Sea has had such authoritative status along with the anti-pollution norms of MARPOL and the new and binding standards of conduct for those operating in the Antarctic.

The other dimension of the typology refers to the relative internationalization of decision-making procedures. The first category of entirely national decision-making would seem only to fit areas where there is no substantive regime or perhaps where national decisions are taken without any involvement in any form of international organization. The other categories are listed by Donnelly (1986: 604), in ascending order:

- International promotion or assistance: institutionalized international promotion of or assistance in the national implementation of international norms.
- International information exchange: obligatory or strongly expected use of international channels to inform other states of one's practice with respect to regime norms.
- International policy coordination: regular and expected use of an international forum to achieve greater coordination of national policies, but no significant review of state practice.
- International monitoring: formal international review of state practice but not authoritative enforcement procedures. Monitoring activities can be further categorized in terms of the powers allowed to monitors to carry out independent investigations and make judgements.
- Authoritative international decision-making: institutionalized, binding decision-making, including generally effective enforcement powers.

All, but perhaps the last, of this hierarchy of decision-making activities can be found somewhere amongst the global commons regimes. The difficulty is that they tend to span more than one of the categories. Thus the Vienna Convention phase of the Ozone regime and the FCCC involved both promotion and information exchange. This in fact would be a good description of the functions of such framework agreements. The Antarctic Regime with the Madrid Protocol, combine elements of both policy coordination through the Consultative Party Meetings and information exchange in terms of 'obligatory or strongly expected use of international channels to inform other states of one's practice with respect to regime norms' (Donnelly, 1986: 604). Most of the regimes under consideration involve international policy coordination, something that well describes ITU procedures to manage frequency and orbit. (The IFRB coordinates national frequency assignments and the Conferences essentially negotiate compromises between the various national and service requirements.) What sets this apart from the 'higher' category of 'international monitoring' is that the latter involves the formal review of state practice. Both the Ozone and Whaling regimes require this although there seems to be a crucial distinction between whether this should involve 'peer review' or some central monitoring body. None of the regimes boast the latter, relying instead on the publication of national reports and investigations. The kind of monitoring activities that Donnelly describes as having independence and a right to make judgements on compliance, hardly exist or exist only in rudimentary form. The best developed, but still flawed, example of an internationally organized attempt to monitor and collect data is

164

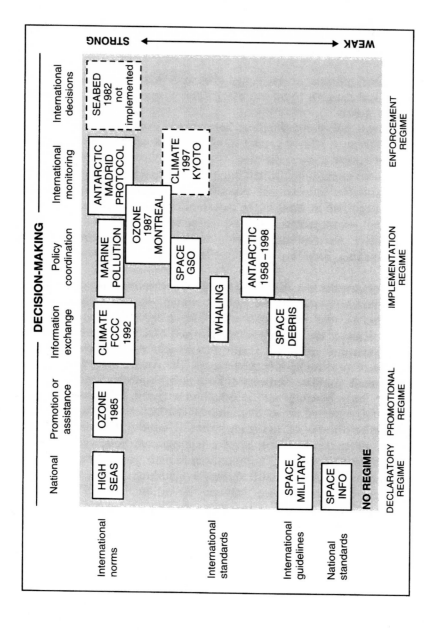

Figure 7.1 A typology of commons regimes (freely adapted from Donnelly, 1996: 603, fig. 1)

provided by the Ozone regime's review procedures. Finally, it is not possible to identify a global commons regime that would fall into Donnelly's highest category of 'authoritative international decision-making ... including generally effective enforcement powers' although the International Seabed Authority as originally constituted might merit consideration.

An attempt is made in Figure 7.1 to locate the various commons regimes in terms of Donnelly's overall typology which extends from no regime or weak declaratory to strong enforcement. There are, first, a number of non-regime areas where purely national standards and national decisions obtain. Included in this category would be failed attempts at regime building in the areas of space-based information flow and, of course the seabed. The regime that was negotiated for the latter might, if it had been implemented, have been placed in the 'strong enforcement' category. Revisions of Part XI of the LoS Convention indicate that a functioning regime for the mining of deep seabed polymetallic nodules will be significantly weaker. The informal understandings on non-interference with military satellites appear to constitute a 'weak declaratory' regime while developments in the international coordination of space debris activity would place this evolving institution somewhere between a 'weak promotional' and 'implementation' regime.

The emergent climate regime can be regarded as 'strong promotional', sharing some characteristics with the ozone regime of the mid-1980s (which might have been classified as 'weak promotional'). It does in fact have significant information exchange rules not found in the Vienna Convention. The full development and implementation of the Kyoto Protocol would promote it to the 'strong implementation' category. The strongest existing commons regimes are of this type. That for stratospheric ozone must qualify alongside the GSO regime (despite some difficulties concerning the availability of reservations). The facility with which the rules of the whaling regime can be avoided would relegate it at best to the 'weak implementation' category while the Antarctic regime modified by the Madrid Protocol, might merit promotion to 'strong implementation'.

What conclusions may be drawn from this, admittedly rough and ready, exercise? First, there are clearly a number of issue areas where regimes either do not exist at all or are weakly developed. They may be contrasted with the regimes that do exist in either the promotional or implementation categories. As far as the global commons are concerned, there are no enforcement regimes. Donnelly's own work on human rights regimes makes particular use of the typology in establishing the shift that has occurred in a number of regimes over the period 1945–85.

> The most striking pattern is the near complete absence of international human rights regimes in 1945 in contrast to the presence of several in all the later periods; ... We can also note the gradual strengthening of most international

human rights regimes over the last thirty years. But even today promotional regimes are the rule, the only exceptions being the regional regimes in Europe and the Americas. (Donnelly, 1986: 633)

He goes on to note that while regime growth may be gradual and incremental in declaratory and promotional regimes, there is a 'profound discontinuity in the emergence of implementation and enforcement activities'. Only regional regimes, which depend upon some form of cultural community, seem to be able to effect the transition.

Global commons regimes, by definition, do not enjoy the support of such cultural homogeneity; although this might be an argument to suggest that many global problems are best treated on the basis of regional cooperation. On the other hand, it is notable from the foregoing brief survey that, at least in comparison with the human rights issue area, it has been possible for implementation regimes of various types to develop. It would be difficult to perform the kind of developmental survey undertaken by Donnelly for our cases, simply because many of the issue areas are very recent, having been thrown up by advances in scientific understanding and technological capability. Even so, it is possible in some areas to observe the strengthening of regimes as they move from the 'promotional' to the 'implementation' category. Such progressive development may be found in the Antarctic, maritime pollution and ozone regimes.

The utility of this type of analysis is that it begins to reveal differences in the various institutions over time and across issue areas. This is clearly of some relevance to the construction of more effective regimes and to understanding the determinants of success. Yet, to reiterate a theme, strength as defined by Donnelly may well be important in a system where the absence of proper central authority is defined as a critical problem, but it is hardly synonymous with effectiveness.

Effectiveness as Behaviour Modification

Observance of rules is the most commonly used measure of regime effectiveness (Haggard & Simmons, 1987: 496; Nollkaemper, 1992: 49; Victor *et al.*, 1998: 6). Governments enter into undertakings to coordinate their behaviour and abide by certain rules, but for these crucial elements of a regime to be meaningful the participants have to be able to 'deliver'. To be effective the regime must serve to modify and sustain behaviour that fulfils its purposes. It is all very well to state this rather obvious truth, but it has proved far from easy to ascertain the extent to which regime norms and rules modify or perhaps merely reflect the perspectives and actual behaviour of governments, let alone their subjects. Within the confines of regime analysis such concerns raise central issues in contention between realists and liberal institutionalists. They can, indeed, be seen as a particular instance of an older debate about the nature and possibilities of international law. Such arguments ought to proceed on the basis of an understanding of the extent to which regime norms and rules are observed in the behaviour of participants. Unfortunately this is a less than straightforward exercise.

Measurement will depend, in the first instance, on whether the regime itself contains rules and standards which are capable of being used as yardsticks. As the UNCED legal survey made plain, 'measurable objectives – such as quantitative targets and technical criteria for compliance – are found in very few agreements' (Sand, 1992: 8). Their availability might, in itself, be regarded as a primary criterion of effectiveness. As Greene (1993a) has pointed out, environmental regimes have not historically been designed, unlike arms control agreements, with monitoring and verification in mind. The pattern is, however, changing as evidenced by the stress on the provision of data and implementation review in both the ozone regime and the FCCC. As with most arms control agreements, the effectiveness of environmental regimes will depend significantly on the extent to which compliance can be assured and national behaviour independently verified and assessed (Greene, 1993a: 175–176). Knowledge about the extent to which countries are meeting their obligations and the existence of verification systems may change behaviour through a number of mechanisms, ranging from deterring non-compliance and building mutual confidence, to promoting learning and technology transfer. In this context, the extent to which commitments are verifiable is a potentially important factor, not only for the analyst wishing to assess formal compliance but more significantly for the effectiveness of the regime in changing behaviour. An instructive example is provided by MARPOL, where standards related to the technical characteristics of tankers and port facilities were more verifiable than indicators of actual oil discharges at sea. As Mitchell (1994) demonstrates, a focus by the MARPOL regime on the former greatly improved its behavioural effectiveness.

Clearly verifiable standards are of little consequence without the availability of reliable data sufficient to allow the evaluation of compliance. Except for the embryonic and as yet unimplemented seabed regime, all the cases surveyed lack an independent central monitoring agency such as, for example, the International Atomic Energy Agency provides for the Nuclear Non Proliferation Regime (although there is provision for limited international on-board inspection in the whaling regime and the CCAMLR). There are good arguments for establishing such a facility for the oversight of the global commons. It will often be necessary to ensure that user rights are established and unimpeded. This already occurs in the Spectrum/Orbit regime through the International Frequency Regulation Board, but this acts rather like a 'claims registry' the main task of which is to ensure that new users do not inconvenience existing users. It is, however, reliant on the national administrations for information. Similar 'claims registries' figured in some developed world proposals in the seabed negotiations and in the defunct Antarctic Minerals Protocol. Such arrangements may be less than adequate when resources are defined as common property and in short supply or when the environment itself is under threat. They may help to police 'claim jumping' but, as with the ITU's orbit/spectrum arrangements, they have an inherent and inequitable bias against latecomers and the international community as a whole unless elaborately constructed along the lines of the Seabed Authority. Experience with the latter and with attempts to provide international observation under the Whaling Convention suggest that full-scale independent international monitoring of the commons is not a practical possibility in the immediate future.

Nevertheless, both at the local and global scale, the restraint in individual behaviour necessary for the sustenance of a commons is usually based on some sort of assurance that others are not taking undue advantage of shared resources. If there is no central monitoring agency the provision of assurance through the observation of compliance will fall to state governments or increasingly, and more informally, to NGOs. Rights of mutual inspection were, for example, written into the original Antarctic Treaty. They have been greatly strengthened by the new environmental impact assessment procedures contained in the Madrid Protocol. Openness and 'transparency', the possibility of keeping one's neighbours under observation, may be difficult to ensure at the international level. This is not just a question of particular states being unwilling to open their societies to external inspection. Some organizations like the ITU, ironically dedicated to the pursuit of the freedom of information, hold their meetings in secret and avoid the publication of conference papers. Elsewhere, in the ozone regime for example, full publication of information may be restricted on account of its commercial confidentiality. Secrecy used to be the working rule of the Antarctic Treaty System, but in recent years there has been an opening up, particularly in response to the work of the NGOs.

The current trend in institutional design is to go a great deal further than establishing a right of inspection. In the new atmospheric regimes, systems for implementation review have been established designed to provide both formal and informal processes where 'participants collect, exchange and review information relating to behaviour and performance' (Greene, 1996: 202). Implementation review systems are not confined to the subsidiary bodies created under the treaties to collect and evaluate data and to manage compliance. The International Institute of Applied Systems Analysis (IIASA) study of implementation (Victor *et al.*, 1998) finds instead that interlocking systems have developed which involve financial institutions like the GEF and the Multilateral Ozone Fund (MLF), *ad hoc* meetings of technical specialists, NGOs and the organs of the European Union. This is relatively new territory and in the stratospheric ozone regime the institutional architecture has grown beyond the expectations of the mid-1980s as a response to the developing requirements of the regime. Such implementation systems do not rely upon a central independent arbiter, but rather, in the manner of a self-governing local commons regime, they ultimately depend upon the provision of information for peer review by the parties. In a newly established regime such as the FCCC the first real test of compliance will be the willingness of parties to provide and share data. Implementation review necessarily functions to provide the kind of mutual assurance against cheating and 'free riding' which is a basic mechanism of CPR regimes (and of arms control verification). It also has other important functions relating to the effectiveness of commons regimes, in terms of strengthening the role of scientific advice, managing compliance and enforcement and playing a significant educative role for the parties (Victor *et al.*, 1998: 51–52).

Non-compliance may arise from the calculated pursuit of self-interest as exemplified by the 'tragedy of the commons' model. The Soviet Union's behaviour in relation to the whaling regime might serve as an example. Yet, it can have a rather more defensive character. For example, a number of European parties to the Montreal Protocol were unwilling to release disagregated data from their chemical producers on grounds of commercial confidentiality.[11] It is just as likely, perhaps more likely, that non-compliance arises from the incapacity of governments. They may simply lack the technical personnel and data gathering facilities to fulfil their obligations under a regime. This problem has been recognized by UNEP and technical and financial assistance with compliance forms a significant part of the ozone regime and the FCCC. Financial transfers in respect of such 'capacity building' are frequently linked to developing country participation in a regime as a form of compensation and once granted they can be utilized as a tool to 'manage compliance' – ultimately through the threat of withdrawal.

It is universally the case that existing regimes assume state jurisdiction and responsibility, whether as a 'flag ' or 'port' state under a maritime regime, an Antarctic Treaty Party exclusively responsible for the behaviour of its nationals, or a 'Notifying Administration' within ITU. Frequently it is assumed that what are often legal fictions, denote actual control. Regimes usually operate at one remove from many of the crucial decisions and behaviour they seek to regulate, trusting in the ability of state authorities to ensure compliance. In the case of European Union members there is a further layer arising from community competence for environmental policy. The private sector bodies that directly use or misuse the commons may often, as in the case of mining, chemical or telecommunications corporations, be organized transnationally. With the best will in the world, national control may be problematic. Thus, the assumption of state responsibility and the inference that commons problems can be handled, if only there is a sufficiently well developed level of international cooperation, begs a large number of questions about domestic regulatory structures and the capabilities of governments. These questions, long neglected in the regime literature, are beginning to be investigated and are clearly of critical significance for the effectiveness of regimes (Jordan, 1998; Victor *et al.*, 1998: Part II).

How well have the global commons regimes performed in terms of effectiveness as behaviour modification? The record is somehat mixed. For some of the more recently established regimes it is difficult to provide much more than informed speculation, but the stratospheric ozone regime is an exception. The regime is one of comparatively few that contain precise performance indicators and it also possesses the most extensive implementation review system. Original targets for the phase out of ozone-depleting substances have been met and exceeded by the main developed country parties. The result has been an actual decline in the consumption of CFCs world-wide from a 1986 figure of 1.1 million tonnes to a 1997 total of 146 000 tonnes (http://www.unep.ch/ozone, 1999). This represents a marked reduction in the developed countries but a continuing increase in consumption in the developed world which is set to decline as developing country parties fulfil their treaty obligations. Meanwhile the developed countries are, in the main, implementing their commitments on the phase-out of HCFCs and methyl bromide. A degree of non-compliance is evident, notably with the failure (admitted in 1996) of some of the 'economies in transition' states (Belarus, Bulgaria, Poland, Russia and Ukraine) to abide by their phase-out commitments under the Montreal Protocol.[12] There is also the problem of illegal trade in banned CFCs which command high prices from those who wish to maintain their existing refrigeration and air-conditioning systems. The trade is estimated to involve 30 000 tonnes of new CFCs per annum in the industrialized countries (http://www.unep.ch/ozone, 1999). State Parties remain in compliance because of their efforts to track, apprehend and punish smugglers.

The maritime pollution regime does include measurable standards and has been operational for a number of years. Nonetheless, there have been problems with mandatory reporting systems. In the MARPOL only 30% of Parties are reckoned to report adequately while the equivalent figure for the London Dumping arrangements is 60%. Independent surveys have, however, concluded that there is substantial evidence of regime effectiveness in terms of real reductions in ship-based pollution and dumping of prohibited materials (Sand, 1992: 163, 156). The GSO regime is characterized by general compliance with the Radio Regulations and Table of Frequency Allocations, something that is assisted by the availability of generous scope for national reservations. The IFRB monitors conformity and superintends the necessary adjustments (coordination) and there are few instances of operating disputes similar to that which occurred prior to the 1979 WARC when the Indian

Administration found itself unable to coordinate a new communications satellite effectively with an existing Intelsat satellite in orbit over the Indian Ocean.

Under the Antarctic regime there has been scattered evidence of non-compliance with a number of specific conservation recommendations, most tellingly reported by NGOs (particularly Greenpeace) rather than through the Treaty procedures. Yet, in the main, the rules for the protection of this extremely fragile environment have not been systematically contravened. The Madrid Protocol provides a much more rigorous compliance monitoring system and the Parties have been implementing its environmental impact procedures in advance of EIF.

As opposed to these relative successes, the whaling regime must serve as a cautionary example of ineffectiveness. This is not only in regard to its very poor historical record and the failure to implement the international observer scheme, but, much more seriously, with recent revelations that one of its most important members (the USSR) cynically disregarded regulations to which it was party but was subject to neither exposure nor penalty. Much attention has been focused on the failure of the regime to modify the behaviour of those in Norway and Japan who remain determined to take whales. They remain, of course, in formal compliance with the regime either through the use of a reservation in the Norwegian case or through the exploitation of the 'scientific' loophole in the Japanese case, although there is evidence of illicit commercial trading in whalemeat.

As for the other regimes it is simply too early to make a judgement. No commercial mining of the deep seabed has occurred but the pioneer prospectors are fulfilling the requirements of the regime. The embryonic regime for space debris, although not codified, appears to be influencing the construction and operation of spacecraft and there is concerted international action in the IADC to evaluate and compare data. Until the Kyoto Protocol enters into force the only formal requirements of the climate regime entail the compilation and review of national greenhouse gas. inventories and programmes and measures. Such activities have been extensively pursued since 1995 and most developed Parties have fulfilled their obligations under the FCCC although there have been some anomalies. The expectation that formal controls for greenhouse gases. will be instituted at the international level has, however, already had a noticeable impact upon national policies. The extensive policy discussions within the EU (European Commission, 1997) and the decision of the Council of Ministers in June 1998 to allocate national shares of the EU emissions reduction agreed at Kyoto provides one notable instance.

A final inescapable question, always raised in discussions of regime compliance and effectiveness, is that of enforcement. It is clearly significant that in drawing up the typology of regimes used in the previous section, Donnelly should have labelled the strongest and best developed 'enforcement regimes'.

The question of enforcement is highly contested and hinges upon the fundamentally opposed assumptions of realists and liberal institutionalists or their counterparts in the international legal profession. An essential point, upon which there is likely to be wide agreement, is that 'vertical enforcement', equivalent to that which is supposedly provided by governments within their sovereign territory, is under current international circumstances highly unlikely:

> The structural realities of international life preclude enforcement by sanctions except in very special circumstances . . . in efforts to induce compliance with international regulation, the prime concern is not to identify and punish 'violators'. Compliance systems in international regulatory regimes are instruments for maintaining a dynamic equilibrium among strongly backed competing interests, so that the regime continues to be viable in a constantly changing international setting. (Chayes & Chayes, 1991: 289)

Thus, a realist would accurately observe that formal enforcement procedures hardly exist in the regimes under consideration except, for example, in the provision of the Montreal Protocol that allows the use of trade sanctions against non-signatories. The inference would be similar to that which might be drawn by the municipal lawyer who is trained to consider the worth of an agreement by the extent to which it is enforceable. International agreements that cannot be enforced, it is argued, are unlikely to be effective, and it is this which leads realists to search for some external power or hegemon able and willing to coerce rule observance.

The response to this might begin with the observation by Brierly (1963: 71–72), originally penned in 1928, that enforcement is irrelevant to the generality of international law:

> For the law is normally observed because . . . the demands that it makes on states are generally not exacting, and on the whole states find it convenient to observe it.

As Henkin's (1979) influential study of international law confirms, widespread, indeed overwhelming, observation of the rules can in large measure be explained by the fact that they will tend to reflect common functional interests. The orbit/spectrum regime provides a good example. It embodies the very practical interests of most members in ensuring that space-based communications systems function properly. Most have few incentives not to comply although, as will be argued in the next chapter, differential stakes in the continuance of such arrangements are of some political significance.

More interesting, in terms of compliance, are areas such as whaling, stratospheric ozone, or maritime pollution where there are immediate incentives for individuals to violate rules that may be in the longer term collective interest. This is, of course, the classic commons problem. At this point it is worth recalling that a key finding of the domestic literature was that effective 'self-management' could exist without the existence of a central authority wielding enforcement powers. This would be in line with the assumption, espoused by institutionalist opponents of realism, that common interests can override individualistic or national interests and that agreements founded on the former may be self-enforcing. But, if the need for enforcement powers is denied, it is vitally important to establish the mechanisms whereby institutions manage to obtain compliance.

Realists would object to any equation of domestic and international experience in this regard. However, as has already been noted, governments cannot always be relied upon to monitor and regulate their subjects. Contrary to the realist assumptions of domestic orderliness and international anarchy, there is substantial evidence that effective regulation by governments may be as problematic at the domestic level as it is at the international. In fact it may be difficult to discuss either in isolation. There are, nonetheless, clear differences between small-scale and a global commons in relation to the compliance problem. For one thing, the small-scale commons is likely to be based on a real rather than an entirely notional community and on continuing personal relationships which will provide powerful inducements to observe collective obligations. These tend to be matters of degree, for the same type of compliance mechanisms may be observed at both levels.

First of all, the force of regime injunctions in themselves should not be disregarded. 'A major influence for observance of international law is the effective acceptance of the law into national life and institutions' (Henkin, 1979: 60). Regimes may not always represent externally imposed constraints on the behaviour of actors with which they have to be forced, or induced in some way, to comply. They may acquire authoritative status and be internalized by governments and indeed the private sector. There are complex and subtle relationships between regimes and policy-making and the educative role of regimes is perhaps more acknowledged than it is understood. This is widely seen as a weakness in the literature of regime analysis.[13]

Nonetheless, the self-interested temptations to violate rules for the sustenance of the commons cannot be avoided. Here the absence of 'vertical enforcement' by a central power may be dealt with, at both domestic and international levels, by various forms of mutual policing or 'horizontal enforcement'. The latter relies on the embarrassment that will be caused to members by the public disclosure of any infractions. Beyond loss of reputation there may be the inherent threat of reciprocal action by other members if rules are broken. There are also instances where individual members (or in the case of the whaling regime (private groups) have imposed their own sanctions against violators.

Reliance on such compliance mechanisms places yet more emphasis on the watchfulness of neighbours or, at the global level, the efficiency of the regime in terms of transparency and monitoring. An important trend revealed in the Antarctic and whaling regimes is that effective exposure to adverse publicity and pressure to comply may be provided by NGOs who do not have to operate under the diplomatic and other constraints that will inhibit state parties.

Although there have been some spectacular cases of rule violation, in so far as it can be accurately ascertained, there has been extensive effectiveness in terms of compliance and behaviour modification. For realists this would not be surprising for rules will tend to reflect the interests of the governments that formulated them in the first place and the strong will in any case be capable of imposing their rules upon the weak. Alternatively, as Gray (1983: 200–201) has argued in relation to the military aspects of three of the commons under consideration:

> States can agree to prohibit or regulate military activity in which they have and foresee, very little interest. Examples are legion, but one may cite the Antarctic Treaty of 1959, the Outer Space Treaty of 1967 and the Seabed Arms Control Treaty of 1971. In all three cases the High Contracting Parties solemnly agreed not to do what they saw no good reason to do anyway.

This cannot really be denied, but it obscures those aspects of the various regimes where serious commercial and even military interests are engaged. Here regimes can provide a powerful external stimulus to live up to obligations that would not otherwise have been observed. Compliance demonstrably involves immediate costs, whether in the use of alternative and more expensive technologies, the abatement of pollution or a reduction in the harvesting of common resources. Despite substantial incentives to cheat or to 'free ride', compliance is achievable through a range of mechanisms other than vertical enforcement – something that is observable at both domestic and international levels.

Questions of compliance and enforcement are essential to the views of regime effectiveness held by both realists and their opponents. For the former the unavailability of credible enforcement explains why regimes *per se* are not likely to be significant determinants of behaviour. Liberal institutionalist approaches, as argued above, flatly contradict this, but the logic of what has been described as the commons problem dictates that effective institutions must be based on some means of assuring participants about the good behaviour of others. It follows that probably the most informative way of treating the compliance problem is to see it as something, not that invalidates the effect of over-ambitious international regimes, but which tends to set the limits of what can be agreed in the first place.[14]

Effectiveness as Problem Solving

If compliance and even behaviour modification is achieved it may not necessarily mean a great deal. For the ultimate test of any regime must be its effects – in this case upon the sustainability and equitable management of the commons. There are real difficulties in assessing this. The situation is not the same as that encountered in the study of the long-standing institutional arrangements for small-scale domestic commons. They have usually evolved to address a well understood collective action problem and the question of effectiveness is not problematic. Incomes are generated, pastures and irrigation systems or fish stocks are sustainably managed. Failures are easily identifiable in terms of overcrowding, rent dissipation or overdrawing from water systems. Berkes (1989) and Ostrom (1990) have no difficulty with effectiveness criteria but devote themselves to the analysis of the different institutional properties of successful and unsuccessful commons regimes. Because most of the areas discussed in the previous four chapters have only recently been considered for collective management, because objectives and scientific assessment are often both disputed and subject to rapid change, and because much of the evidence that would be required to arrive at conclusions about impact simply does not exist, the situation with the global commons is very different.

Regime objectives at the level of norms and principles have often proved to be plastic. Witness the transformation of the purposes of the whaling regime from exploitation to strict conservation or the changing and expanding objectives of the Antarctic regime. Adaptive capacity and an ability to reformulate objectives in the light of evolving scientific understanding – what is known in the jargon of the FCCC as the 'review of commitments' – is now seen as a significant element of institutional effectiveness (Greene, 1996: 209). Perhaps sustainable development can provide a yardstick? This idea of the 'integration of environment and development objectives' has, as the UNCED legal survey notes, 'figured prominently as a common denominator in authoritative statements issued in the context of many different agreements and instruments' (Sand, 1992: 8–9). It cannot, however, provide some consensual external standard by which regime effectiveness may be judged. As the whaling example shows it may, on occasion, be part of a dispute, in this instance between the advocates of sustainable harvesting (including the concept's author Gro Harlem Brundtland) and those who adhere to preservationist values. More fundamentally, and much of the international politics of UNCED bore witness to this, it optimistically yokes together what may be contradictory aspirations. Efficient management of the commons may not accord with the achievement of equity and development – of 'common heritage principles'.

Neither may the stated objectives in the legal documentation of the various regimes provide much guidance. As the UNCED legal survey found, these are formulated in 'highly general and abstract terms, which makes it difficult to evaluate goal achievement' (Sand, 1992: 8). The ozone regime, with its specific objectives, provides in this, as in other ways, an exception. The whaling regime no longer has quotas which are above the catch levels that hunters were able to achieve, but the various loopholes in the moratorium and continuing relative ignorance about cetacean populations makes assessment difficult. The preamble to the FCCC represents an object lesson in the diplomatic art of accommodating seemingly incompatible positions. But the Convention also contains a non-binding bench-mark for the reduction of carbon dioxide emissions, later developed into specific Kyoto targets, against which the performance of the developed nations will in future be assessed.

In the end, if some independent criteria for the assessment of effectiveness are required, it will be necessary to step back from the regimes themselves and specify the problem. Yet even establishing that a problem exists can involve scientific, economic and political controversy. The question of greenhouse-gas-induced climate change is the most dramatic example of a 'contested' problem and there are still those who argue that the global warming hypothesis is fundamentally flawed and the FCCC an economically damaging irrelevance (IEA, 1994). In the other issue areas there is more consensus as to the existence and nature of the problem although there will be differences as to its extent and the need for regulation. For example, many of those involved in the use of the frequency spectrum argue that there is really no problem of shortage that cannot be easily solved through technological innovation.

A focus on global commons implies a particular sort of problem involving the efficient and sustainable management of shared resources. While there may be wide agreement on this, questions of equity have equal prominence. Striking a balance between the two broaches the political and economic divisions at the heart of the 1992 UNCED. There is no consensus on the meaning of sustainable development and all that can be done is to rate the effectiveness of the various regimes in terms of both environmental and management as well as equity and 'common heritage' criteria.

The basis for such evaluation will often be provided through the efficient performance of the knowledge generating function associated with commons regimes. In this the ozone regime is relatively well served. The Technology and Economic Assessment Panel reported in late 1991 that the production and consumption of ozone-depleting substances was subject to a more rapid decline than that required by the Protocol (Tolba & El-Kholy, 1992: 56–57). Nonetheless, simulation studies indicate that, even under optimistic assumptions about a complete ban on all human sources of ozone depletion, it will take until around 2050 until the ozone layer is restored to pre-CFC levels. The long lifetimes of CFCs mean that even though emissions are sharply decreasing, stratospheric concentrations are still increasing (http//www.unep.ch/ozone, 1999). In terms of equity, the regime has, under the London Revisions, begun to deal with the compensation and technology transfer issue.

The orbit/spectrum regime allows technically efficient utilization of a common resource; scores of satellites operate on a continuous basis in GSO. Yet, as the outcome of events in 1985–88 demonstrates, it has hardly been modified to 'guarantee in practice' equitable access. All the figures for the ownership and control of satellites and occupation and use of GSO show that is very heavily skewed in favour of the developed countries despite its enormous potential for LDCs.

In terms of meeting the rather narrowly specified objectives of the Dumping and MARPOL Conventions, the maritime regime has fared well. Dumping without permit has been much reduced, but there has on the other hand been only a slight decrease in tonnages actually dumped.[15] Intentional oil spillages have been reduced from 1.47 million tonnes in 1981 to 568 000 tonnes in 1991. As Mitchell (1993: 236–240) demonstrates there is some difficulty in arriving at the sources and extent of marine pollution, partly on account of the lack of adequate monitoring. Very much the same can be said of the whaling regime, which has always suffered from a lack of reliable data and indeed basic understanding. Here the situation in terms of sustaining great whale stocks appeared to be desperate prior to the moratorium and its effects are unclear. (Advocates of renewed whaling argue that minke whale stocks are now very substantial.) Antarctica, enjoying 'more comprehensive international agreements for environmental protection than any other part of the earth's surface (Tolba & El-Kholy, 1992: 778), has, according to the general consensus of opinion, been effectively preserved. The arrangements covering this global commons, however, also protect it from the unwanted intrusion of 'common heritage' principles and of the uninvited majority in the UN system.

Having specified a problem, and especially one that appears to be being contained or ameliorated, there remains the question of the role of regime in its solution. As the argument in Chapter 2 suggested, the inherent nature of commons problems suggests that regimes are likely to fulfil certain types of function if they are to be effective. But this does not allow the observer to draw conclusions about the impact of particular regimes. In some ways this would pose unanswerable questions or at least ones that require sophisticated guesswork and counterfactual analysis. What would have happened if a particular regime had not been in place, or if it had been differently constructed?

Discussions of institutional effectiveness often neglect the characteristics of the issue areas and problems that they 'govern'. There is a tendency to rate institutions as more or less effective, when it is the complexity, difficulty and extent of the problem that is probably the key determinant. Underdal (1994) has usefully conceptualized the determinants of regime effectiveness in terms of the inherent benignity of the problem itself on the one hand and the 'problem solving capacity' of regime institutions on the other. Problems differ in their degree of difficulty, and while some may require little in the way of institutional development, others call out for very substantial innovation and regulation if they are to be tackled at all.

To use Donnelly's terminology, some commons problems may be effectively treated by 'promotional' regimes, while others may need strong 'implementation' or even enforcement regimes. An example of an 'easy' (or even 'non' problem) is provided by Antarctic mineral extraction. It requires no great exercise of imagination to see how the whole political and institutional situation would change were it to be the case that mining became a realistic commercial proposition. The ozone issue area seems to be characterized by a relatively soluble problem although, as with the other cases, the five rule-related functions of commons management (specified in Chapter 2) had to be fulfilled. The degree of difficulty tends to rise when the problem is defined in terms of equity and redistribution and 'common heritage' principles are embodied in the regime. The problems with institutionalizing seabed mining or equitable access to orbit provide illustrative examples. In the case of the seabed, mining was a more feasible proposition than in the Antarctic. The fact that it became less of an immediate likelihood as the negotiations proceeded, did not markedly ease the difficulty of negotiating common heritage arrangements.[16] This is not to say that such matters should, or even could, be avoided if effective and sustainable management of the commons is to be achieved. In the longer term institutions themselves are unlikely to be sustainable if they do not address development and equity issues.

A conception of global problems shared by practitioners of the various natural science disciplines, ecology and green politics is that they must be understood holistically. The natural systems which constitute the commons are indivisible and intertwined. It is often argued that they must be treated as such if they are to be sustained. This represents, in effect, a call to widen the scope of an issue area and, if heeded, will probably increase the extent and complexity of the 'problem' as defined by the regime. It may not necessarily increase the difficulties encountered. The 'ecosystem' approach of the CCAMLR which recognizes that individual fishery stocks cannot be conserved in isolation represents an increase in scope and complexity as compared with existing fisheries regimes but is widely regarded as being potentially effective in terms of the conservation of Southern Ocean resources. However, there has been very widespread criticism of the FCCC because of its initially narrow approach to the problem of climate change. Although the problem has been defined in terms of 'sources and sinks', the focus has been upon emissions reductions by the developed states rather than the preservation of forest sinks. There are good political but not ecological reasons for this. One might go even further and argue that a truly effective regime to ensure the mitigation of climate change would have to embrace measures to stabilize and reduce human population. Another view, stressing equity and the legitimate aspirations of the developing world, would point out the necessity both for North–South resource transfers and the radical alteration of the international and monetary and trade regimes if climate change is to be averted (Middleton *et al.*, 1993). Almost the only point of agreement would be that a regime mainly devoted to achieving limited reductions in developed world greenhouse gas emissions is inadequate.

Ineffectiveness in terms of a 'fragmented' institutional response to complex and interrelated problems is of course a consequence of the way in which issue areas are constructed.[17] It is also the result of the sheer interconnected complexity of problems such as marine pollution. While it may be possible to target specific 'intentional' polluting activities by ships, there is a recognition of the vast scale and multiplicity of sources of the land-based pollutants that are responsible for most of the degradation of the ocean commons. An implication of this is that the 'effectiveness' of the ozone regime stems mainly from the confined and relatively simple nature of the problem that it addresses, rather than the inherent qualities of the institutions created on the basis of the Vienna Convention.

Although there is a natural and justifiable tendency to treat the mismatch between the scope of a regime and the dimensions of the underlying problem, perceived ecologically, as a source of ineffectiveness, the criticism can equally be regarded as impractical. There may be strong political reasons for avoiding the creation of all encompassing regimes, however logically attractive they may appear. The epic attempt at codification involved in the Third Law of the Sea Conference is often now regarded as a precedent to be avoided. The fact that the parties were 'conscious that the problems of ocean space are closely interrelated and need to be considered as a whole' (UNCLOS, 1982, preamble) did not induce a sufficient number of them to ratify the agreement until over a decade after signature. The failure of most developed countries to ratify at all, even when most parts of the Convention were acceptable, was because they were conjoined with the controversial seabed regime in Part XI. It may even be asserted that global problems are most efficiently treated by a series of regional regimes, which include smaller and more manageable numbers of participants and may profit from a degree of cultural homogeneity. However, in the cases under consideration it is difficult to see how they could be anything more than local components of some global arrangement.[18]

Regime Assessment

Regime effectiveness has been conceived of in a number of ways. Logically in the final analysis all are subsidiary to environmental and other outcomes. Yet these may be impossible to assess. Establishing the relationship between internal architecture, the centralization of authority, the degree of compliance and the actual solution of commons problems is far from easy.

First there are a number of issue areas where commons problems have been identified, but where regimes have not been constructed or implemented. There is no regime governing space-based information flow despite attempts to create one based on 'prior consent'. Although an institution based on common heritage ideas with a sophisticated internal structure and extensive powers was negotiated for the deep seabed it was not implemented and seems likely to acquire life under different circumstances and in a significantly weakened form. The deep seabed remained, at least until November 1994, technically without governance but neither has commercial activity occurred. There are some rudimentary and declaratory understandings relating to the military use of the space commons, which have nonetheless proved effective in the sense that anti-satellite weapons have been tested but not deployed.

A global regime for whaling has existed since 1946, but on virtually all criteria for most of its existence it must be judged ineffective. It did include the whaling nations but failed dramatically in terms of scientific, regulatory and monitoring functions, having very little effect in terms of the preservation of stocks of great whales. Unfortunately, even since the transformation of its norms and principles during the 1980s, weaknesses remain and impact is uncertain.

The Antarctic regime has a number of institutional shortcomings, in terms of the legal force of its rules and the effectiveness of its monitoring, but by general consent it has proved effective in preventing the despoliation of the continent if not in espousing common heritage concepts. Since 1959 it has demonstrated a significant capacity for change in response both to scientific findings and new international concerns at the level of norms and principles. It may be added that many of the problems it seeks to solve are not yet difficult in the sense that it is regulating for future contingencies where commercial interests are not, at present, heavily engaged.

The orbit/spectrum arrangements demonstrate extensive institutional development and constitute an 'implementation' regime. There can be little doubt that it has provided an effective system of governance for the complex business of coordinating and operating hundreds of GSO satellites. There may be some query as to its future ability to cope with technological change and a very major query as to its effectiveness in treating outer space as a common heritage and ensuring access to orbit on an equitable basis.

The components of the regime for marine (ship-based) pollution are similarly well developed, have a good record in terms of compliance and, within narrowly defined objectives, have had substantial impact. The difficulty that arises is that the pre-eminent sources of marine pollution lie elsewhere and are subject to no effective regulation.

On most of the criteria discussed in this chapter, the ozone regime may be deemed effective. Yet its full impact, as measured by the restoration of the stratospheric ozone layer, can only be assessed at some point in the latter part of the twenty-first century. Its apparent institutional effectiveness may not provide a perfect model for the emerging climate regime. The reason is very simple and derives from the enormous difference in the relative difficulty of the two problems. No amount of 'effective' institutionalization can compensate for that.

Notes

1. Smith (1993: 44) has argued that such 'optimism and suggesting solutions dominates the writing' with the consequence that it will be easy for the unsympathetic to marginalize the whole enterprise.
2. As examples of such adaptive technical organizations and institutions Strange cites the ITU, IMO and WMO.
3. There have been various attempts to explore the meaning of regime effectiveness. Oran Young (1994: 140–160), for example, suggests six meanings of effectiveness as: problem solving, goal attainment, behavioural effectiveness, process effectiveness, constitutive effectiveness and evaluative effectiveness. Process effectiveness is similar to my category of 'effectiveness as international law' and behavioural effectiveness equates to 'effectiveness as behaviour modification'.
4. For a comprehensive treatment see Birnie & Boyle (1992), while Sands (1993) provides a series of essays on the development of international law. On the question of compliance see the important collection edited by Cameron *et al.* (1996).
5. An intriguing proposal for trusteeship is suggested by Stone (1993) who addresses the commons problem from the standpoint that the interests of the commons have no established legal advocate. In the same way that trustees are appointed to represent the interests of minors in domestic law, so 'guardians' of the commons should be appointed at the international level.
6. See for example Codding's (1981) study of influence within the ITU that provides an empirical demonstration of these participation problems.
7. On the participation of NGOs within the UN system see Willetts (1996).
8. At the Eleventh Special Consultative Meeting in Madrid in 1990 the Antarctic and Southern Ocean Coalition (comprising the main interested NGOs) plus bodies such as the IUCN and international organizations such as the WMO had observer status. There is now a fairly standard clause in current treaties which allows observer status as of right for non-member states and international organizations, and also for NGOs unless one-third of the state members object. (See, for example, Art. 7.6 of the FCCC.)
9. In the case of the 1992 ITU Constitution and Convention there was a new provision under which after the ratification deadline in 1994 the new instruments would enter into force between the ratifying Members (Art. 58(1)). Those failing to ratify after a further two years are deprived of voting rights (Art. 52(2)).
10. Matthew Paterson in *International Affairs*, 69, 3, July 1993, p. 583.
11. A solution was found through the Commission of the EU which employed the private accountants KPMG Peat Marwick to conduct a confidential analysis (Greene, 1998: 106–107).
12. Compliance should be assured by 2000 if sufficient international assistance is forthcoming from the GEF and from a special World Bank initiative under which

a group of the major developed countries will buy out production facilities in the Russian Federation (UNEP/IUC website, 1999).

13. 'This neglect has extended to the issue of how regimes actually influence national policy choices, a question closely related to the issue of compliance and regime strength' (Haggard & Simmons, 1987: 513). The concern is echoed in another survey of writing on international cooperation by Milner (1992) which points to the very substantial development of international system-level thinking about regimes as contrasted with the paucity of concern with interactions spanning the international/national divide and the role of domestic political processes. Recent work has begun to address this problem (for example, Schreurs & Economy (1997)).

Establishing the relationship between norms and behaviour empirically poses a number of problems. Formal rules may merely represent existing governmental practice and there may be a need to erect counterfactual estimates of what might have occurred in the absence of the regime. As Goertz & Diehl (1992) point out in their attempt to provide the means for research into this relationship, it may be possible to establish variations between established norms and patterns of behaviour – to isolate non-compliance – but it is almost impossible to establish compliance, in effect to demonstrate that a norm is responsible for particular behaviour. At the same time:

> the measurement of norms must include some indication that the strength of the norm varies according to the actor it is supposed to influence: that is, its strength at the international level may not be reflective of how widely accepted it is in the state whose behaviour is supposed to be affected. (Goertz & Diehl, 1992: 646)

An approach to the problem is provided by Haas et al. (1993) who place major emphasis on regime effectiveness in terms of 'increasing governmental concern' and 'building national capacity'. The third of their 'paths to regime effectiveness' is the 'enhancement of the contractual environment', a key element of which is compliance monitoring. The IIASA study (Victor et al., 1998) makes the first comparative survey of the domestic impact of regimes in terms of implementation. It stresses the extent of public participation, the role of target groups and the patterns of access to decision-making. It also highlights the special difficulties of the economies in transition.

14. This relationship has been explored by Downs et al. (1996) in their article entitled 'Is the good news about compliance good news about cooperation?'.

15. From 7.2 to 5.8 million tonnes in the period 1976–86. Source: Sand (1992: 185).

16. Arvid Pardo (1984: 74) himself has pointed out that the three basic assumptions of the seabed regime did not correspond to reality:

> deep sea-bed mining will comprise at least the harvesting of polymetallic sulphides and cobalt crusts and probably also mineral rich muds, in addition to manganese nodules; deep sea-bed mining will take place not only in the international seabed area, but also and predominantly within the expanded legal continental shelf . . . Hence, the International Authority will enjoy no monopoly and will be obliged to develop the minerals of the international seabed area in difficult competition with a number of coastal states.

17. UNEP has long noted and attempted to remedy the fragmentation of international efforts to treat commons problems by sponsoring interorganizational coordination through Designated Officials for Environmental Matters (DOEMs) and through the development of the grandiosely named System Wide Medium Term

Environmental Plan (SWMTEP). (See Tolba & El-Kholy, 1992: 760–772 and Imber, 1993.)

18. In terms of environmental protection, by far the greatest number of agreements exist at a regional level. Less than half of the 124 international environmental agreements mentioned in the UNCED (Sand, 1992) survey allow global membership. In the area of marine resources and pollution 9 agreements with global scope are mentioned alongside 27 regional agreements which include all fisheries agreements other than the CCAMLR and the ICRW. This, in the main, reflects the fact that the problems concerned have an essentially regional scope although linked to the global level. It may also be evidence of a tendency found by Donnelly (1986) that the smaller number of participants and cultural homogeneity that may be present at a regional level is more conducive to the negotiation of 'strong' regimes.

8

Explaining Regime Incidence and Change

The uneven incidence and effectiveness of regimes suggests that there is no simple link between the existence of a 'collective action problem' and the emergence of an adaptive institution to cope with it. For some of the global commons there are well developed 'strong implementation regimes' but in others such institutions hardly exist, or are too fragmented to be effective. A central question for the various theories about regimes is how such variations are to be explained. Why, for example, is the regime for orbit and spectrum well established whereas there are no effective arrangements to 'govern' some of the potentially damaging uses of earth-orbiting satellites? We might also ask about the ways in which the specific characteristics of regimes are determined.

The main preoccupation of regime theorists has been with questions of stability and change. Regimes may exhibit a high degree of institutional inertia; they would hardly fulfil their function of providing a stable and long-term set of expectations as to behaviour, if they did not exhibit some stability. Although a great deal of academic attention has been devoted to regime change, stability in the face of the erosion of the terms of interdependence which determined their creation is perhaps more remarkable. The maritime regime of 'freedom of the seas', which lasted from the time of Grotius to the mid-twentieth century, is a case in point. It survived massive alterations in international economic and power structures and technological change which, even by the end of the nineteenth century, had made definition of the territorial sea in terms of the range of a cannon shot seem more than anachronistic. For 40 years the whaling regime was saddled with a fundamental concept, the 'blue whale unit', which was widely recognized to be inadequate in terms of conservation even at the time of its invention in the mid-1930s. Although born of technological change the ITU frequency regime has been systematically outpaced by advances in technological capability. The waves of deregulation and globalization that have transformed the telecommunications industry have inspired recent, possibly belated, attempts at wholesale re-structuring. Without some degree of stability there can be no regime, yet institutional inertia is also a major factor accounting for regime ineffective-ness in the face of rapid change in an issue area.

On the other hand, the global commons cases also provide examples of very rapid and significant regime change (as defined in terms of alteration in norms and principles). During the 1980s the whaling regime was transformed from a system for harvesting to one designed to ensure its almost complete avoidance. The Antarctic regime evolved slowly but at the end of the same decade a significant change of direction had occurred with the replacement of CRAMRA by the Madrid Protocol. It is an encouraging trend that, contrary to experience with the Bretton Woods institutions (always a central concern for regime analysts), the development of global commons regimes has usually been in the direction of greater effectiveness – with respect to curbing environ-mentally damaging behaviour. A proper understanding of long-term regime effectiveness also requires that such patterns of stability and change should be adequately explained.

Of particular interest and importance has been the way in which 'new' commons issues, particularly those relating to the atmosphere, appear on the international agenda. Whereas the older issue areas and related regimes (for the high seas and frequency use) evolved, sometimes spontaneously over a long period, recent awareness of global environmental change has focused attention on building regimes through negotiation. There is a very practical need to understand, not just the negotiation processes, but the ways in which agendas are set and the circumstances in which public support and pressure on governments and corporations is mobilized. Some caution is in order here, for what may appear at first sight to be an exercise in regime creation may, in fact, be another instance of regime change. Nonetheless, the kind of explana-tion that may be satisfactory in accounting for the long-run rise and fall of a regime may not serve to illuminate the much more proximate patterns of causation that determine the specific timing and content of regime creation or negotiated change.

The literature on regime creation and change has been dominated by two types of explanation, structural and utilitarian. As discussed in Chapters 1 and 2, both are significantly connected to parallel explanations of 'domestic' commons institutions. Structural explanations are provided within both the realist and Marxist or structuralist traditions. Utilitarian explanations, stressing self-interested cooperation, reflect one form of liberal thinking. This chapter cannot provide anything like a full review of even these two dominant approaches. What it will endeavour to demonstrate, in preliminary fashion, are the ways in which they might be applied and the extent to which they seem capable of accounting for the patterns of incidence and change that have been observed in our set of cases. It will be argued that neither provides a fully comprehensive explanation of regime creation and change. Although preceding from what are normally seen as antagonistic premises they share a state-centric myopia towards domestic and transnational politics. This is compounded by a neglect of the shifts in public values and scientific knowledge which are of no little significance in the contemporary history of the global commons. More recent work, focusing on cognitive change and sub-state and transnational politics has begun, as discussed at the end of the chapter, to rectify this deficiency.

Structural Explanations

Structural explanations of regime formation and change relate the possibilities of cooperation to the structure of the international system and more specifically to the existence of a hegemon or dominant power. There are 'variants of hegemonic theory to suit most political tastes' (Strange, 1987: 557). All are 'structural' theories but the key distinction is in the structures to which they refer. Reference to structure alone is unenlightening for this overworked word merely denotes the relatively permanent pattern of relationships between the elements of a system (Vogler, 1996). Neo-realists, Marxists and others, who would accept neither label, employ the concept and indeed the notion of hegemony, but there are fundamental differences in the structures that are relevant to their divergent explanations.

For neo-realists, following Waltz (1979), the key structure is the decentralized and anarchic pattern of power relationships between sovereign states. From a perspective inspired by the Marxist tradition, the relevant structures denote the historical relationships between productive forces and social formations. The material dominance of the centres of capitalism (the UK in the nineteenth century and the USA in the twentieth) can, indeed, be outlasted by their legal and ideological manifestations. The approach developed by Strange (1988) differs from Marxism by envisaging four equal and interacting structures including security and production structures but adding finance and knowledge as well. An actor occupying a dominant position in any of them will enjoy structural power, but an actor like the United States which continues to occupy such a position within all four will be a hegemon. Although this contradicts American concern with a post-1971 'loss of hegemony' (Strange, 1987), the one point on which all structuralists would concur is that in the post-1945 period the only conceivable candidate for a hegemonic position, in terms of any of the conceptions of international structure that have been discussed, is the United States.

Structural Power

Writing from a Marxist perspective Cox (1983: 353) speaks of hegemony in terms of a 'fit between power, ideas and institutions'. A consideration of the principles and norms of the global commons regimes provides some evidence of this kind of hegemony, specifically in the way in which the domestic legislation of the dominant state (representing the most advanced form of capitalism or the state occupying a dominant position in Strange's four structures) may be almost effortlessly reflected in the make-up of international institutions. This is true structural as opposed to relational power. The existence of a regime and its particular characteristics can be explained as a manifestation of hegemony. The hegemon's norms, principles and rules will serve to coordinate behaviour in the entire system.

A revealing commentary on what this can actually mean comes from a decidedly non-structuralist source, a technical work on the development of space telecommunications law:

> I note that the law of England as applied to maritime matters through the Admiralty Court had significant impact and became for most purposes international maritime law . . . History, therefore leads me to suspect that US Law may become international space law . . . By reason of its technical skills, its domestic market and its entrepreneurial attitudes, the USA is a major leader in space matters. That 'lead' may result in much US Law becoming the language in which problems are discussed and solved. (Lyall, 1989: 419)

This view is confirmed by study of the evolution of more general telecommunications regulation (Hills, 1989).[1] It is not confined to high technology areas. In a number of our cases it is possible to trace the origins of regime norms to US domestic legislation. The enclosure of the ocean commons by 200 mile EEZs, which was such a feature of the Third LoS Convention was first suggested by a unilateral American proclamation by President Truman in 1945. In the next year the United States led the creation of the ICRW, and discussion at the Washington Conference was based on its draft. However, contrary to the US line in many other issue areas, development could not be associated with the interests of corporate America (there was no remaining commercial interest in whaling). Rather, the upsurge of environmentalist pressure group activity and legislation within the United States had very direct effects upon the regime in that US policy was driven by its burgeoning domestic legislation – notably the 1972 Marine Mammals Protection Act.[2] A primary stimulus to the creation of the ozone regime was the 1977 US domestic legislation banning CFC use in aerosols. This in effect produced something of a common interest between environmental groups and the US chemical industry in the international extension of these restrictions.

That the domestic legislation of a state occupying a dominant position in security, financial, knowledge and production structures should colour the norms and rules of international regimes should not be surprising. It represents the most effective and perhaps insidious form of structural power. However, it cannot be asserted that such hegemonic influence will always occur. In the period of British naval supremacy the early development of the regime for the radio spectrum actually ran counter to British ideas and the attempt (inspired by the Admiralty) to establish a Marconi monopoly of ship-to-shore radio. The United States was unhappy with the innovations in GSO planning during the 1980s. The very emergence of 'common heritage' ideas in so many of the relevant issue areas illustrates the limits of hegemony. American and Western ideas and legal concepts were not automatically incorporated. They were, as we have seen, enthusiastically challenged, but with rather limited practical results. Perhaps these are the exceptions that demonstrate an otherwise pervasive structural power.

The other way in which the Marxist or historical materialist tradition relates to the politics of managing the global commons, is that, while the relevance of institutions *per se* will be denied, the political contradictions within them may be subject to a robust and historical explanation. There are a number of ways to approach this but all are founded on the conception that the degradation of the commons and the antagonisms between users are driven by global capitalism's 'underlying processes of production and accumulation' (Saurin, 1996: 95). Marxist approaches mount a significant challenge to the whole enterprise of international cooperation (the concluding chapter will return to this point) but they also have a more proximate utility. For example, Paterson's study of the politics of the FCCC concludes that while liberal-institutionalist analyses provide the most convincing account of the specifics of interstate cooperation, Marxism 'explains the depth of the North–South rift over global warming in a way that no other perspective can' (Paterson, 1996a: 179).

Hegemonic Stability

The other, and most widely discussed, structural approach is represented by variants of the hegemonic stability thesis. They are generally rooted in realist assumptions about the impossibility of self-interested cooperation and the requirement that order be imposed. The Hobbesian assumptions underpinning realism are shared by a number of scholars who have considered solutions to the 'tragedy of the commons'.

> if ruin is to be avoided in a crowded world, people must be responsive to a coercive force outside their individual psyches, a 'Leviathan', to use Hobbes' term. (Hardin, 1978: 314)

The realist notion of a hegemon in the international system is thus closely related to the idea that small-scale domestic commons cannot be effectively governed without the intervention of an external power to create and enforce rules. The hegemon constitutes the functional equivalent of central government at the international level. An important version of the thesis, propounded by Kindelberger and others, specifically outlines the role of the hegemon as the solution of the otherwise intractable problem of the supply of public goods.[3]

The type of hegemony envisaged is significantly different from that understood by Marxists and other 'structuralists'. It will be proactive, exercising a leadership role in the creation and maintenance of regimes.[4] The role of the hegemon can even be benevolent in that, in the public goods version of the thesis, the small and weak will derive disproportionate benefits from collective provision. To enter into an hegemonic role with this in mind seems decidedly at odds with normal realist assumptions and writers such as Gilpin and Krasner are evidently more at home with an exploitative hegemon. It may also be that the concept of hegemon as public goods provider represents in itself an intellectual manifestation of hegemony (Tooze, 1990).

Whether the hegemon's purposes are the provision of collective welfare or simple self-aggrandisement it will, by definition, have such material and military strength as to yield a dominant position in the international power structure.[5] Regime formation and maintenance will require hegemonic leadership providing stability and order (or public goods) for other states in the system. The prediction that such regimes will atrophy as the hegemonic power itself declines appears as a logical consequence of the thesis.

Hegemonic stability has the virtue of providing a clear hypothesis. Although it might be possible to associate the evolution of the rules for the high seas with British and then US hegemony, this would hardly provide a plausible explanation of the actual negotiation of maritime rules under UNCLOS III. Benedick (1991) has argued the case for active US leadership in the face of European recalcitrance in the creation of the ozone regime. This has been disputed by other participants and tends to neglect the initial hostility of the Reagan administration and the negative attitude of its successor to the creation of the multilateral ozone fund in 1990. The negotiations for a Convention on Climate Change provide a further instance of the absence of hegemonic leadership.[6] While a plausible case for such leadership might be constructed for the ozone regime, US policy has consistently resisted and only grudgingly accepted, in watered down form, the institutionalization of a climate change regime. The 'perceived costs (both economic and political) of action were very much higher and encouraged the US in particular to stand firm' (Brenton, 1994: 194). European calls for targets and timetables prior to Rio were rejected and the FCCC was only saved by 'eleventh hour' diplomacy which agreed a text shorn of binding commitments and which President Bush felt able to sign in an election year construction. The United States continued to find itself cast in such a defensive role by the strength of domestic opposition to cuts in energy use throughout the later development of the FCCC. At Kyoto it was once again a question of whether the US would assent at all to the proposed greenhouse gas. reduction commitments and more than a year after the eventual compromise agreement there was still no American ratification.

This long-drawn-out episode is illustrative of the negative or veto role of a 'hegemonic' state. Whereas it may be difficult to assert the necessity of hegemonic leadership, it is far more plausible to suggest that hegemonic acquiescence is required. Without wishing to enter into the debate on 'lost hegemony' it is indisputable that the scale of the US economy and its military and diplomatic capabilities make it difficult to envisage a global regime that could function without at least tacit American approval. The fate of the seabed mining regime and the Third LoS Convention after the change of course by the incoming Reagan administration provides a salutary lesson in this regard. However, as Porter & Brown (1991) demonstrate, the veto role is not limited to an assumed hegemon. Depending on the issues and the degree of commitment involved, a variety of actors may be in a position to prevent the establishment of an effective regime and indeed to derive bargaining leverage from this understanding.

Oligarchy

The hegemonic stability thesis has never seemed entirely at home in the canon of realist thought. Classic realist writings have been predicated on the anarchic nature of the interstate system and on an historically based understanding that bids for hegemony have been met by successful countervailing coalitions of states. Hobbesian assumptions, shared with many domestic theorists, led to the conclusion that it was precisely the unlikelihood of the emergence of a 'Leviathan' that differentiated international from domestic politics. Thus, balance of power rather than hegemony has been regarded as the necessary basis of international order. This would point towards an explanation of international cooperation that relied upon the mutual accommodation of the interests of a group of powerful states rather than the leadership of a hegemon. The outcome of such accommodation might be described in terms of oligarchy – rule of the few – rather than hegemony. As realist writers such as Krasner (1985) have pointed out, a successful regime must reflect underlying power capabilities. This view appears inherently plausible. In general the effective commons regimes that have been reviewed usually include the major military and economic actors in the system.

Some, like the Antarctic Treaty regime, are exclusive or have developed voting rules that ensure that the interests of major players with a tangible economic interest in the decision-making of the regime cannot be overlooked. Arguably, this rests upon a similar foundation to the permanent member veto in the UN Security Council – the understanding that if business is to be carried on the most powerful actors must at least acquiesce. It is difficult to imagine a frequency and orbit regime that would operate if it were not supported by the United States, European Union, Japan and Russia. The same can be said for the ozone regime where initial negotiations were dominated by OECD countries and the old USSR. Krasner (1985) makes the point effectively in his comparison of the fortunes of the LoS seabed mining regime and the Antarctic Treaty regime. The latter is seen to be stable and successful because a limited membership of powerful states, possessing real interests in the region, have monopolized decision-making. One might add that during the 1980s the regime was rather adroitly expanded to include additional countries like China which might have had sufficient political weight to cause problems if left outside the exclusive circle of consultative parties. Part XI of the LoS Convention was never likely to provide a stable seabed regime because support for it was 'incongruent' with underlying power capabilities (Krasner, 1985: 263). With the achievement of 60 ratifications in 1993, serious attention was devoted to negotiating a modification of Part XI such as to allow the United States and its developed world partners to subscribe to the whole Convention.

Issue-related Power

'Underlying power capabilities' are not always self-evident. Nor are they necessarily the same for every issue area that may concern us. If there is a single power structure that governs relationships across a range of issues and if it is possible to generalize about state interests, then a realist approach may well provide a powerful explanation of the incidence of regimes. It would simply be necessary to determine the identity of the actors occupying dominant positions in the power structure and to isolate those areas in which their interests were coincident. A relevant example is provided by the implicit regime covering anti-satellite weapons (ASATs) and the military uses of outer space during the Cold War. There were only two superpowers in military competition, and it would have been easy to predict that if each had plausible reasons to refrain from the deployment of ASATs then limited cooperation and a regime reflecting these interests would be put in place. It might also be possible to advance a similar explanation for the Antarctic Treaty System. The protagonists in the Cold War had a mutual interest in institutionalizing the avoidance of military confrontation in this area.

However, it is now generally acknowledged that with the widening range of issues on the international political agenda and the growth of interdependence, simple realist explanation of the type displayed above is unlikely to have universal applicability – if indeed it ever did. Long before the additional complexity produced by the ending of the Cold War, scholars such as Morse (1976) and Keohane & Nye (1977) were arguing that it was no longer possible to discern a single power structure based upon relative military capability. Military force was simply unusable or wholly inappropriate to a range of issue areas that reflected the tightening web of interdependence between societies. Many of the global commons reflect such interdependence, whether in the use of the radio frequency spectrum and GSO or in the awareness of shared vulnerability to the effects of stratospheric ozone depletion or the build up of greenhouse gases. Traditional power relations, according to Porter & Brown (1991: 108), 'have no direct impact on the outcomes of specific environmental conflicts'. Under conditions of 'complex interdependence', argued Keohane & Nye (1977), one could no longer expect changes in regimes to reflect the overall politico-military power structure. Even dominant actors such as the United States would have difficulty in 'linking' various issues together in order to bring their position in the general international 'pecking order' to bear. Instead power would be 'issue specific' and related to capability within a particular area.

It follows that, in the politics of the global commons, power configurations will tend to be issue specific. Because of the close association between economic, resource and communications, maritime and environmental agendas it is likely that the dominant set of players will be drawn from the OECD states. However, there are important variations. The Soviet Union as a major space power was a significant force in negotiations on frequency and orbit as were India and China and, to a certain extent, Brazil. As Codding (1981) demonstrates in his empirical study of the 1979 WARC, influence may relate not just to the size of a country's economy and its stake in the issues under discussion, but also to the scale and technical qualifications of the delegations participating in the regime's decision-making procedures. In the Antarctic Treaty System the structure is limited and defined and gives an important role to countries such as Chile and Argentina.

Although the issue-related power configuration for the ozone negotiations would seem essentially to involve an oligarchy of the main CFC producers and users (the EC, USA and to a lesser extent the Soviet Union and Japan), the important subsequent role of China and India should not be neglected. The 1990 London negotiations on the Interim Ozone Fund were dominated by the leverage exerted by these two states, both of which had the potential to disrupt the Montreal Protocol by proceeding with plans to manufacture ozone-depleting chemicals outside the agreement at the same time as they were to be phased-out in the developed North. The definition of the climate change issue area and the subsequent negotiations in the INC necessarily relied upon agreement amongst the main carbon dioxide emitting countries and were in the end subject to the veto of the United States. While an agreement, however weakened, was at least achieved, the parallel attempts to draft a Forests Convention failed dismally. Here the powerful 'issue-related' position of Malaysia as the world's largest exporter of tropical timber was significant.

However, in a number of instances, regime change occurred in ways which appear to run counter to predictions based upon issue-related power. The whaling regime provides an interesting and unusual example of a set of actors taking decisions on the future of the industry, in which most no longer have a direct commercial interest. The Japanese and Norwegians, who do, have witnessed the transformation of the regime in ways inimical to these interests. The negotiation of the 1988 WARC-ORB agreement and indeed the serious discussion of 'common heritage' ideas are similarly inexplicable in terms of the exercise of issue-specific power.

Interdependence, Linkage and Regime Politics

The politics of the whaling and other regimes suggest that an assessment based on a narrow view of issue-related power is not convincing. For Norwegian and Japanese societies, whaling has symbolic and culinary significance but it hardly constitutes a major sector of the economy. A key element in the success of NGO campaigns against commercial whaling was the creation of issue linkage whereby the pro-whaling nations understood that failure to accept the moratorium would place much more significant fishing industry interests at risk in terms of consumer boycotts (and even state action).

When regimes reflect extensive interdependence they have an evident value for powerful users. This provides a potential source of leverage for actors which do not rank highly in terms of issue-related power or their position in the overall power structure. If regimes rely on formal organizations then it is possible for their norms and principles to be challenged by voting coalitions of the weak (always assuming that 'one state one vote' voting rules pertain). Facility in operating within the networks and organizations associated with regimes can in certain circumstances prove a significant political resource in preventing or promoting change.[7]

Since the 1970s it has been the strategy of the Group of 77 and the Non-Aligned to mobilize majority coalitions in organizations behind NIEO or NWICO objectives. The implicit strength of such a grouping was observed during the LoS negotiations which maximized the impact by tying a whole range of issues together in one package. In the event the United States and its allies were able to avoid such pressure by rejecting the Treaty *in toto* and taking desirable elements such as the 200 mile EEZ (also a high priority of many members of the G77) on an *à la carte* customary law basis. Initial perceptions of Western vulnerability in terms of the 'free passage' aspects of the regime were re-evaluated with the conclusion that there was no compelling reason, in terms of real interdependence, to proceed with the Convention if it continued to include the objectionable Part XI.

By contrast, the orbit/spectrum regime did govern an area of extensive interdependence in which a potential disruption of the rules would have major consequences for powerful advanced countries. As one US commentator wrote at the time of the 1979 WARC:

> Economically, US dependence on telecommunications discourages ignoring ITU because of the destructive economic chaos that would ensue. More important, US leadership in telecommunications technology promises substantial economic benefits from a global expansion of communication. (Branscomb, 1979: 142–143)

It was the implicit threat to the regime posed by Third World plenary majorities which may explain why Northern countries were prepared to take demands for 'equity in orbit' seriously even if they remained convinced that they made no economic or technological sense. The result was the two WARC-ORB Conferences of 1985 and 1988 and the plan for guaranteed national GSO slots. It should be cautioned that this largely symbolic agreement involved only marginal change in the orbit/spectrum regime and represented a time-consuming irritation to developed countries, rather than a major threat to their interests. Nevertheless, it demonstrates that leverage by an LDC coalition is possible using organizational strategies, if the regime is regarded as sufficiently important by powerful members. This is in clear contrast to the parallel events at UNESCO, where the UK and US were happy to leave the organization. There was, during the period of the 'equity in orbit' debates, only one disregarded suggestion that ITU was equally dispensable.[8]

Despite the general failure of NIEO to reshape many of the rules governing the global commons and the international political economy as a whole, the recent development of atmospheric regimes holds out new sources of potential leverage. These may be based to an extent upon an ability to use 'organizational' strategies but much more significantly on new sources of issue-based power. The case of the 1990 London Conference and the Interim Ozone Fund has already been remarked upon. Existing signatories of the Montreal Protocol were inherently vulnerable to potential CFC producers operating outside the newly created ozone regime. Indian and Chinese delegates were able to argue at the 1990 Conference that they could hardly be expected to adhere to the rules if it meant that their nascent CFC-based industries would be disadvantaged and the costs of national development raised by the need to buy in new and expensive Northern produced ozone-friendly products. Thus the cost for the developed countries of a workable regime (i.e. one including India and China) would be special treatment on the phasing-out of CFCs, compensatory funding and technology transfers. The US Government, which had approached the London meeting with a policy of firm resistance to any compensatory funding, was persuaded to agree an Interim Ozone Fund for technology transfers as the price of Indian and Chinese adherence to the regime.

North–South antagonism has been a central theme in the development of the climate change regime. Relationships between participants have, however, been rather more complex than this label would suggest. Amongst the OECD countries there has been a clear division between JUSSCANZ and the EU. The former grouping has tended to advocate minimal 'targets and timetables' for carbon dioxide reductions while the latter has regarded itself as leading the way in making reductions. The oil-producing states of OPEC – normally associated with G77 positions -- have been the most hostile to taking any measures whatsoever that might damage their energy exports. On the other hand, the AOSIS states, also members of G77 have, for reasons of national survival, pressed for maximum levels of commitment to greenhouse gas reduction. The conflict within G77 came to a head at the Berlin CoP in 1995, with a decision of the majority to back reductions by the OECD states and the negotiation of the Kyoto Protocol. The further development of the regime will turn upon the politics of interdependence. In the immediate term the US Senate (Byrd Resolution) has made ratification of Kyoto contingent upon the adoption of 'voluntary' commitments to greenhouse gas. reductions by developing countries. On the other hand, members of the G77 have long-standing demands for additional funding and capacity building as the price of their adherence to the regime without greenhouse gas. reduction commitments. In the longer term, the projected growth of the Chinese and other developing economies ensures that there can be no effective response to predicted climate change without their full involvement.

Utilitarian Explanations

The debate on regime formation and change is sometimes represented as a new form of the classic dialogue between realism and liberal-idealism. This is misleading, for the 'liberal' side of the argument has been dominated by utilitarian and 'institutionalist' thinking. (It will be argued below that this relative neglect of other liberal approaches, whether idealist, pluralist or functionalist, is unfortunate.) Liberal-institutionalism is firmly grounded in concepts of self-interest and utility shared with neo-classical economics. In the latter's abstract world there was for many years little concern with the role of institutions. Self-interested actors maximized their utility and the operation of Adam Smith's 'hidden hand' ensured the achievement of an equilibrium providing the optimal distribution of scarce resources. (The different assumptions underlying realism or neo-mercantilism are evident here. Gains are assumed to be absolute, while in the realist conception of competition for status, power and wealth gains are always seen as relative to those of other players.) In classical liberal theory mercantilist interventions by governments or attempts to rig markets were regarded as deeply undesirable for they disrupted the underlying harmony of free economic intercourse between self-interested actors. As Keohane (1984) has acutely observed, in a world assumed to be in harmony there is no need for cooperation.

However, political economists have long been aware that there are occasions when the application of pure free market principles may lead to outcomes which are anything but optimal. The commons, as discussed in Chapter 1, provide one instance of such 'market failure'. Where the 'solution' of enclosure or privatization is unavailable or where 'public goods' such as clean air must be provided, state intervention is justified at the domestic level. At the international level equivalent governance functions necessitate regimes, whether hegemonic or otherwise.

Such an economistic appreciation of the need for institutions actually represents one piece of common ground between neo-realist regime theorists and liberal-institutionalists. The essential difference between them is over the manner in which regimes arise and are sustained. As we have seen, the orthodox neo-realist position stresses the need for a 'quasi governmental' hegemon. The alternative liberal-utilitarian view is put by Young (1989: 199) who, significantly, claims that there is a consensus on the importance of institutions:

> By now, everyone is aware that rational egoists operating in the absence of effective rules or social consensus often fail to realize feasible joint gains and end up with outcomes that are suboptimal (sometimes dramatically suboptimal) for all concerned . . . It follows that individual actors frequently experience powerful incentives to accept behavioral constraints of the sort associated with institutional arrangements in order to maximize their own long-term gains, regardless of their attitude to the common good.

It is noteworthy that cooperation is self-interested and does not require a hegemon, neither need it be based on anything as fickle as notions of the 'common good'. The best known development of the first point is provided by Keohane's (1984) attempt to account for the persistence of economic regimes 'after hegemony'. His argument was explicitly based upon a microeconomic view of the functions of institutions. Regimes are created and sustained because they deal with problems of market failure at the international level, improving the quality of information and providing a set of stable expectations which will enhance the possibilities of cooperation. Much of the argument rests on the way in which institutions serve to minimize transaction costs while at the same time allowing compliance monitoring and 'decentralized enforcement' (Keohane, 1984: 245).

Keohane's arguments are directly relevant to the collective action problems identified by Hume and Olson. They do not, however, get to the heart of the utilitarian rationale for self-interested cooperation in the management of commons. This must be interdependence amongst users. Commons institutions coordinate the avoidance of mutually damaging behaviour, a function which can in the case of common sinks also be represented as the provision of public goods. The utilitarian argument is that the key to explanation of the emergence and pattern of development of regimes for the global commons is to be found, not so much in the mutual gains that become available through cooperation in an institutional setting, but in the management of mutual vulnerability. Essentially, we may expect strong and well developed regimes in issue areas where actors are aware of extensive vulnerability interdependence. This fundamental proposition is supported by empirical evidence gleaned from the systematic study of small-scale common property resource regimes. The most important reason for participants to change rules to achieve joint welfare is a shared 'common judgement that they will be harmed if they do not adopt an alternative rule' (Ostrom, 1990: 211).[9]

The proposition that strong and effective regimes will tend to exist in issue areas that exhibit high levels of vulnerability interdependence can be supported by reference to the global commons cases. The orbit/spectrum regime fulfils many of the criteria of effectiveness and does reflect very high levels of interdependence in the utilization of space-based communications. The significant uses of orbit and spectrum for military and civilian purposes in which vast investments are at stake and upon which many activities are critically dependent, could not occur efficiently without a complex regime. The disorder and interference that would occur in the absence of these precisely specified 'expectations of the behaviour of others' is unthinkable. This is a very different situation to the one that pertained in another space commons issue area – involving attempts to erect a regime for 'prior consent' to direct broadcast by satellite (DBS) transmissions. The existing absence of regulation or 'free flow' regime served the interests of the United States and a number of European governments and corporations. The situation was complex, particularly with attempts to regulate at the European level, but the underlying truth remained that there were few incentives in terms of mutual harm arising from interdependence sufficient to override ideological and commercial objections to imposing restraints on DBS.

Maritime pollution and resource conservation regimes have also been built upon shared conceptions of mutual vulnerability. There has been growing awareness, reflected in regime construction, that all users will be harmed if individuals are allowed to pollute the sea or over-exploit resources. Rules based upon such understandings have been erected under MARPOL and are evident in the ecosystemic approach of CCAMLR. Whereas, in the radio spectrum area, vulnerability interdependence in terms of signal interference seems to have been accepted from the beginning, this has hardly been the case with the depletable resources of the global commons. With whales and fishery stocks exploiters have either been ignorant of the consequences of their actions or have applied high discount rates to the future, confident that the apparently limitless resources of the sea would provide for alternative catches. Mutual international awareness of the finite nature of maritime resources and long-term vulnerability interdependence has been somewhat belated.

There is still a real contrast between the regulation of these maritime activities and the prolonged failure to implement the painstakingly negotiated seabed minerals regime. In some ways this has analogies to the question of 'prior consent' rules. The optimal situation for developed world mining consortia would be one in which there was minimum regulation and they lobbied hard for this during the LoS negotiations. There were no mutual vulnerability incentives in relation to seabed mining *per se* to suggest the need to accept the restrictions and 'common heritage' aspects of the proposed regime. The fact that the Carter administration and other Western governments were prepared to negotiate at all derived, as we have seen, from linkage to another set of issues.

Neither do the relationships between members of the IWC approximate to the conditions of interdependence between users of scarce but renewable resource that one would expect, under utilitarian assumptions, to find in an effective commons regime. Once again, the mutual vulnerabilities are not really comprehensible in terms of a strictly utilitarian approach. In the non-whaling countries policy was driven by public distaste while, for the whalers, vulnerability was associated with linkage to significant trade and fisheries interests. In the Antarctic case it would also be difficult to argue a strictly utilitarian case. The regime rests upon the lack of realistic interest in the commercial exploitation of the commons. All members have, by the activity criterion, some interest in the continuance of their scientific research and may genuinely support the view that the preservation of Antarctica is a question of long-term common fate interdependence, although this cannot be claimed as a high national priority. Arguably, most would see that their interests were vulnerable to an unrestrained territorial and commercial scramble for the continent. The effort that has been put into protecting the current ATS regime from interference in UN forums provides some measure of the importance that members attach to it.

By contrast, the establishment and development of the ozone regime is certainly explicable in terms of interdependence. Mounting evidence of the scale of ozone layer depletion and public awareness of the medical implications whereby humanity as a whole was vulnerable to continued CFC production and use, created the context in which governments came to negotiate.

Deriving from this were more specific commercial considerations of mutual vulnerability. Since the 1977 aerosol ban in the United States (followed by similar action in a number of other countries) both Du Pont and its counterparts were under mounting pressure to develop 'ozone-friendly' CFC alternatives. 'These would be costly to develop and higher priced than CFCs. So the makers had an interest in an early and clear framework of regulation applying to as many potential competitors as possible' (*Economist*, 16 June, 1990: 22). Creating a regime that would provide a stable framework of business expectations was thus of some importance to the chemical industry and the various governments. There was mutual vulnerability in the sense that, without an effective regime, those engaged in the development of CFC alternatives (with an estimated value of $100 billion) would face the risk of being undercut by the continued manufacture of cheaper CFC-based products elsewhere. Such considerations were at the heart of the bargain that created and sustained an effective and rapidly developed and implemented ozone regime.

The question of an effective climate regime is much more problematic. It is not at all clear that even if the projections of anthropogenically induced global warming enjoy the kind of widespread acceptance achieved by earlier studies of stratospheric ozone layer depletion, a sufficiently widespread awareness of mutual vulnerability interdependence would emerge. The projected temperature and sea level changes, unlike ozone layer depletion, will occur well into the next century and far beyond the normal time horizons employed by governments and even business corporations. Neither is it immediately apparent that all the signatories of the FCCC would experience mutual harm and vulnerability to changes under the 'business as usual scenario' for, as Skolnikoff (1990: 81) observes, 'not all the effects will necessarily be damaging, some activities will be enhanced and some localities and nations will benefit by the changes, at least in relation to others'. The asymmetries between the parties in their assessments of costs and benefits related to global warming and the measures proposed for its curtailment are already complex and difficult. Although there may, on the grand scale, be a conception of global common fate interdependence, it cannot be said with any certainty that the rather specific and practical forms of interdependence that have sustained effective commons regimes will pertain in this case.

Plural Interests and Values

Both realists and utilitarian-liberals have pitched their explanations at the level of state actors, accounting for their behaviour in terms of fixed assumptions about interests and motivation. Classic realist conceptions of the national interest, involving the territorial security and legitimacy of the state and the endless struggle for prestige and relative power advantage, have some application to governmental behaviour in relation to the global commons. They cannot be divorced from the exigencies of 'high politics'. This was always the case in discussions of the Law of the Sea, where the extent of control exercised over territorial seas and the rights of passage through straits were of clear strategic importance to maritime powers. Thus, the views of the United States Navy and Department of Defense were of particular importance in shaping policy toward UNCLOS III. An analogous situation exists with the regimes for the GSO regime. Substantial parts of the frequency spectrum used for military purposes, particularly at X band for satellite communications, are sanctified under ITU rules but by tacit agreement are not a proper subject for discussion at Radio Conferences. The original basis of the entire Antarctic Treaty System was an understanding related to the freezing of territorial claims and demilitarized status. The politics of territoriality dominated much of UNCLOS III, with the enclosure of vast tracts of ocean and seabed within the 200 mile EEZs.

Foreign policy towards the commons may not, nevertheless, simply be an exercise in the mercantilist pursuit of territory or resources. A corrective to such views and a restatement of realist concern with the essentials of state prestige and legitimacy is provided by Krasner (1985). The thesis of his *Structural Conflict* is that the campaign for a NIEO and its manifestation in the seabed negotiations did not really concern economic and resource interests and inequalities. Rather, NIEO is to be understood as an essentially political campaign having the important function of raising the status and increasing the legitimacy of weak newly independent states. Thus, arrangements for the Seabed Authority and Enterprise had a politically symbolic rather than a genuine wealth distributing function. A similar line of argument might be employed to understand the equity in orbit campaign. Although some of the leaders such as India had a specific national interest in frequency allocation arising from the practical need to plan new satellite systems, the overwhelming majority of LDC participants did not and probably could never hope to. However, obtaining a national satellite orbital position under the new 1988 plan which had technical parameters not inferior to those of other countries became a matter of status, resolutely pursued by many administrations. The significant function of regimes in the maintenance of state autonomy receives extensive support from research into the institutions governing international transport and communications (Zacher with Sutton, 1996).

Whether the interests pursued are strategic, political or material the important point for realists is that they are national and essentially in conflict. Thus negotiations over common pool resources are regarded as a competitive struggle for a limited stock of fish, orbital positions or whatever. Such a Hobbesian view has great difficulty in explaining more enlightened and cooperative behaviour, particularly in the absence of an hegemonic power.

This critique of realism is hardly novel. Writers in the liberal tradition, broadly construed, pointed to the emergence of 'complex interdependence', the 'dissolution of the issue hierarchy' and the boundaries between what used to be called 'high' and 'low' politics – the latter denoting issues of technical cooperation and welfare (Keohane & Nye, 1977). Such matters, which are very prominent in the politics of the global commons, had, it was argued, a significance equal to that of traditional national interests. The conception was of a system in which, contrary to the simplicities of orthodox realism, a plurality of interests and types of actor engaged in a political process transcending national boundaries.

Such a model seems particularly appropriate to the politics of the global commons and to the explanation of the formation of particular agendas and the specific life histories and characteristics of regimes. It is therefore somewhat surprising that much liberal writing on the theory of regimes has been not only utilitarian but state centric. Eager to pursue 'parsimonious' and economistic lines of explanation many theorists have tended to forsake the pluralistic world of complex interdependence and concentrate on utility maximizing nation state players.

The behaviour of actors in the politics of the global commons (including non-state actors) cannot be fully accounted for in terms of the fixed motivational assumptions of either realism or liberal institutionalism. There is, in particular, a need to recognize a normative dimension to political action and perhaps to refer back to often maligned traditions of idealism and liberal internationalism.

Clearly there are some episodes in the history of the global commons that cannot be explained in terms of either realist or utilitarian behavioural assumptions. The obvious example is provided by the transformation of the whaling regime, motivated by an influential set of campaigning NGOs which mobilized developed world public opinion behind a decidedly non-utilitarian concept of the rights of cetaceans. A further recent and significant instance is provided by the changes in the London Convention outlawing the disposal of low-level radioactive waste at sea. This change ran contrary to the specific interests of a number of significant developed states (including the USA and UK) and was not supported by scientific evidence of probable harm. The critical determinant appears to have been NGO pressure and public opinion.[10]

Environmental activism to preserve the global commons combines universalistic moral concern with a conception of a collective human interest. Concepts such as intergenerational equity in relation to climate change only make sense within such a non-utilitarian framework. It is easy to be cynical about the moralism of green politics, wilderness values or non-anthropocentric conceptions of rights, but as a real factors in the deliberations of governments and other actors they cannot be discounted. The whaling moratorium, the extension of the Antarctic Treaty in the Madrid Protocol and indeed the emergence of global atmospheric issues and the Rio process itself stand as testaments to the significance of 'ethical' public opinion. For some this represents a paradigm shift and a deep normative change. Perhaps this may, to an extent, represent wishful thinking because of the widely shared estimate that only such a transition from materialism can sustain the political action necessary to safeguard the planet. Environmental politics is seen as 'an indicator of a major value shift in modern society. It is one of a number of social issues that characterize the transition from modern to postmodern society' (Caldwell, 1990: 101). The mechanics of this new form of idealist politics are beginning to receive attention in the academic literature. Wapner (1996), for example, has conceptualized 'world civic politics' in terms of non-state-orientated political practices that exist alongside the state system. Here the focus of attention has been analysis of the role of the environmental NGOs (Princen & Finger, 1994; Wapner, 1996).

World civic politics is more significant for some issue areas than for others. The long-running campaign for a whaling moratorium has already been mentioned as the 'flagship' of environmental NGO activism. Aroused public opinion was a highly significant factor in maintaining the pressure for the creation of the ozone regime. Elsewhere, regime politics relating to the util-ization of GSO or the seabed does not excite widespread interest. Trans-national activism and public opinion also appear to move in a cyclical fashion related both to 'shocks', such as the Chernobyl disaster, the discovery of the Antarctic ozone hole and unusual weather conditions, as well as to the com-parative strength of competing issues, notably those relating to economic performance. The period from the mid-1980s through to 1992 appears to have been particularly propitious in terms of the political impact of environment-alism. Such factors, which may involve deep but incalculable normative shifts, hardly figure in conventional theories of regime development and change. Yet they clearly enter the calculations of democratically elected politicians, having a direct effect upon the composition of agendas and the possibilities of international agreement. The 'conversion' in 1988 of Mrs Thatcher, if not to green politics, at least to an appreciation of the salience of environmental issues, provides an example.

A close analysis of the way in which agendas have been set and regimes changed will frequently lead back to domestic politics, or perhaps more accurately the complex intermeshing of domestic and international politics. Although environmentalist campaigns have had significance, they are but one amongst a range of competing forces and interest groups. It might be argued that, in this essentially domestic competition, environmental values only prosper when they are not directly opposed by more powerful strategic or economic welfare concerns. This would be the case for whaling and Antarctica, but it would also highlight the inherent problems of developing an effective climate regime where short-term economic welfare is sometimes regarded as being fundamentally at odds with longer term environmental responsibility. Despite the shifts that may have taken place in public environmental consciousness, economic growth still remains the lodestar of most politicians. A graphic illustration was provided by the debates within the European Community on energy taxation prior to UNCED and President Bush's refusal to sign the original draft of the FCCC on the grounds that in an election year he would not 'trade American jobs and prosperity' for the uncertain benefits of a binding commitment to emissions targets. Notions of sustainable development and 'win win' solutions are attempts to manage this conundrum (Gray & Rivkin, 1991). Perhaps the success of the ozone regime is founded upon the way in which the preservation of the stratospheric ozone layer was associated with actions that had relatively low economic costs and, indeed, promised new commercial opportunities.

The best documented case revealing the complexities of internal politics concerns US policy and the Law of the Sea. Although analysis of the various LoS regimes will refer to the interests and policy of governments:

> When one looks beneath the label 'US Government' to discover the agency and interest group actors that determine national policy, the source of discontinuity in policy becomes intelligible. A host of incompatible and self-interested coastal and distant-water concerns are vying constantly to determine national policy. On any given set of issues one side may prevail; then the other may win the next policy victory. (Hollick, 1981: 374)

The bureaucratic battles that accompanied the negotiation at UNCLOS III and the ultimate reversal of US policy under Reagan were actually cited as a precedent by those involved in similar domestic combat over the desirability of developing an ozone regime (Benedick, 1991: 46). In this instance the lobbying by US environmental NGOs supported by elements of the US Government assumed transnational dimensions, leading to complaints to the State Department from the British Government (Benedick, 1991: 39). US climate policy has been subject to equally convoluted bureaucratic infighting where the interests of rival agencies and indeed the opinions of strategically placed individuals have loomed large (Hopgood, 1998).

For a number of reasons – the open and pluralistic nature of its political system and its assumed hegemonic role – most detailed analysis of the domestic and bureaucratic politics of agenda setting and regime change has concentrated on the United States. It would be possible to uncover such detail for the other global commons cases and for at least some of the other major countries involved. A precise answer as to why a set of issues were constructed in a particular way or a regime negotiated to a specific outcome might require such depth. Accounts by participants usually display an awareness of a political process which is complicated; transcends the boundaries of national and international systems; and is subject to the vagaries of individual commitment and ability, public opinion, bureaucratic politics and the lobbying of special-interest groups. At present this hardly amounts to a coherent explanation of regime development – more a caveat to the standard utilitarian approaches. The importance of such 'intermestic' relationships is as clear in this area as it is in the connections between international and domestic legal systems. They probably represent a key area for future theoretical and empirical investigation. By contrast, theories of cognitive change and the role of epistemic communities have already been presented as a significant alternative or at least adjunct to established approaches.

Changing Cognitions – Epistemic Communities

Functionalism, the notion that harmonious international relations and even the integration of separate nations could best be achieved by the promotion of non-controversial technical or 'functional' cooperation across national frontiers, has been a significant strand of twentieth century liberal thought in International Relations (Mitrany, 1975). Arguably the study of epistemic communities and an approach to regime creation and change that emphasizes their role in altering actor cognitions is its latest manifestation.[11] An epistemic community is:

> a network of professionals with recognized expertise and competence in a particular domain and an authoritative claim to policy relevant knowledge within the domain or issue area. (Haas, 1992a: 3)

Epistemic communities may have a transnational character, where members share certain principles and causal beliefs and, most significantly, are 'engaged in a common policy enterprise' (Haas, 1992a: 3). They are likely to have a special relevance for commons and environmental regimes which, as we have seen, are unusually dependent upon the generation of scientific knowledge. Specific issues of regulation frequently have a highly technical character (orbit and frequency use or the specifics of oil tanker operations for example) which render them incomprehensible to any but a restricted group of specialists. Thus it is possible to identify 'communities' of specialists dealing with the atmospheric chemistry of ozone layer depletion, maritime pollution, space telecommunications engineering and Antarctic science. These groupings are often informal and rely on professional contacts, but they also have an organizational basis in, for example, SCAR, ICES, the IWC and its Scientific Committee, the consultative and technical committees of the ITU, and the IADC. A very dense network of international scientific programmes and organizations has grown up, often with overlapping memberships and fulfilling a knowledge creation function for many of the regimes under consideration.

Some indication of the extent of the network and its interconnections in relation to the umbrella issue of climate change is provided by the diagram of inputs to the IPCC process given in Figure 6.1. There was a large and increasingly influential epistemic community of climate change scientists and supporters in international organizations during the 1980s. The creation of the IPCC in 1988 represented a successful attempt by governments (notably the US government) to assert national control over the climate change research process. Its title was not a misnomer, for it was to be an Inter-governmental Panel (Boehmer-Christiansen, 1993: 380). Epistemic communities are not exclusive, for in the analysis provided by Haas (1992b) and others they include 'like-minded' politicians and officials; Richard Benedick, head of the US delegation at the Montreal Protocol negotiations is, for example, treated as a member of the ozone epistemic community.

The assumption is that epistemic communities can play a decisive role in regime formation and change by altering the cognitions of actors and providing authoritative consensual knowledge to uncertain policy-makers. It is claimed that a range of environmental treaties (including the Montreal Protocol) negotiated under the auspices of UNEP were 'concluded through the influence of epistemic communities' (Haas, 1990a: 349). In the climate change case an extensive epistemic community appears to have been very significant in establishing global warming as an issue during the 1980s but far less influential once the intergovernmental negotiations for a convention were set in train (Paterson, 1996a: 147).

Although a strict line of descent from neo-functionalism is denied, epistemic explanations clearly do have much in common with this older view of international organization.[12] In particular there is the concept of the opposition between ecologically sound consensual technical knowledge and politics. It is not too far-fetched to characterize the epistemic community as a virtuous transnational technocracy with a mission to educate and lead short-sighted national politicians into the paths of environmental righteousness. An example of such thinking, which is widespread, is provided by Johnston's comments on the maritime pollution regime, where effectiveness depends 'above all' on the interaction of the relevant technical elites and on their ability to 'depoliticise the process'. In this he would wish to back 'los tecnicos' in their constant if understated struggle with 'los politicos' (Johnston, 1988: 205).[13]

This concept of the distinction between objective, rational scientific knowledge and politics is a familiar and contested one. It should be treated with appropriate scepticism. For many years within ITU an engineering subculture, which had at its core the notion that the function of the organization was merely to provide rational and politically neutral solutions to technical problems, held sway. The observations of Third World and other critics that this institutionalized a set of frequency use arrangements, heavily skewed towards the interests of the developed countries, were greeted as a deeply unfortunate attempt to 'politicize' the technical work of the organization. Litfin's (1994) study of the discourses surrounding the creation of the strato- spheric ozone regime provides a subtle demonstration not of the influence of scientific/technical knowledge but the interpenetrative relationship between power and knowledge.

The function of technical consensus may thus be to circumscribe political choice in what many would regard as undesirable as well as desirable ways. Nor can the members of epistemic communities be immune from evidently political considerations. The most obvious are that national scientists funded by national research agencies will represent national interests. The members of the original Working Groups II and III of IPCC, concerned with impacts and policy, were inevitably criticized for this. However, the most important 'political' concerns to be found within supposedly 'apolitical' epistemic com- munities are more parochial. It should not be forgotten that the scientific establishment is itself an interest group with careers and research budgets to protect, just as the bureaux of the various scientific and functional organ- izations will compete for the scientific and policy 'turf'. The detailed analysis of the creation of the climate change issue area by Boehmer-Christiansen (1993, 1996) provides ample evidence of the interplay of special interests within a very extended epistemic community.

Because epistemic communities are relatively amorphous and because their claim to significance is based upon the engineering of cognitive change amongst policy-makers, their real impact is difficult to establish. The strongest case is made out by Haas (1992a) for the transnational community which was united in urging the adoption of measures to preserve the stratospheric ozone layer. It clearly occupied influential positions within national administrations, set much of the agenda and influenced the timing and stringency of regime rules. The key to this success was the peculiar dependency of regime creation on the evolving atmospheric science and above all the pressure generated by the consensual nature of that knowledge. This contrasts with the increasing conflicts over the validity of climate change predictions related to the FCCC regime. The impact of epistemic communities appears to have been marginal in the case of the whaling regime, where cetologists were divided on key scientific issues (Peterson, 1992) and in the maritime pollution regime where no research programme existed and scientific investigation had a 'low salience' (Mitchell, 1993: 236). Such conclusions are supported by Porter & Brown (1991: 24–25), who describe the essentially limited role of epistemic commu- nities, especially in the whaling and maritime pollution (radioactive dumping) issue areas. There have been similar problems with the generation of scientific consensus to support the development of the CCAMLR rules, and the Madrid revision of the Antarctic Treaty. As regards the latter, a close student of the regime concludes that:

the scientific commmunity did not function as an epistemic community (in policy advocacy) nor even successfully in support of its own interests in the regime. (Elliot, 1994: 195)

Recent work on epistemic communities, thus, affords them an unrealistically privileged position in explanations of regime creation and change. It may be more accurate to regard them as another interest group or lobby in the often complex transnational and transgovernmental politics that surrounds some of the recent efforts at regime creation. Clearly, their influence will vary with the extent to which consensual scientific knowledge is sought by other powerful players and the extent to which that knowledge serves to advance and legitimize other interests. Thus, for example, Boehmer-Christiansen (1993, 1996) provides an analysis of the origins of the climate change issue area based upon the political activities of energy interests and national governments with a stake in the regulation to reduce carbon dioxide emissions. They had good reason to support and publicize the work of the climate scientists. Neither is it excessively cynical to suggest that the immediate beneficiaries of the FCCC are the research communities themselves for, if it did nothing else, the Convention pointed towards a continued and expanded effort in research and data gathering.

A Synthesis?

Regime theorizing inevitably derives from a conflicting set of traditions: realist, liberal and to a lesser extent Marxist. It may range across different levels of analysis, as is evident in the differences between the state-centric assumptions of liberal-utilitarianism and the domestic and transnational emphasis of liberal-pluralism. This may give rise to the view that the varying lines of explanation surveyed in this chapter are incommensurable – incapable by their very nature of being combined in any meaningful way. Rather, as in the case of the general philosophical debate over the commons, the observer is invited to choose between Hobbesian and liberal accounts which take up diametrically opposed positions on the possibility of cooperation. They rest upon axioms concerning the wellsprings of human behaviour. On the realist/ Hobbesian side the drive to maximize power, status and autonomy with resultant conflictual behaviour is assumed. Significantly for the management of the commons, individual gains will always be assessed and valued in relation to the gains of other actors. In contrast, neo-liberal thought proceeds from the assumption that individuals (and indeed state governments) are rational utility calculators who will be persuaded, under the appropriate circumstances, to so order their affairs as to provide joint gains and the most efficient allocation of resources. Marxist structuralists and cognitive theorists operate from similarly incompatible premises.

Is the analyst logically condemned to plump for one or the other or can there be some form of synthesis? The path towards such a synthesis is opened up if one is prepared to accept the undogmatic view that each approach captures a significant facet of human behaviour, more or less relevant to the interpretation of regimes. It is greatly facilitated if there is an understanding that we are not dealing with one central question, for which there are rival explanations, but with a variety of questions. One approach may well be greatly superior to another when the evidence on the long-term incidence of regimes is assessed. It may, however, have very little explanatory power if we are interested in the precise terms under which regimes are negotiated or the processes of change. Thus, Young (1994: 115) has argued that neither the neo-realist or neo-liberal explanations of the general incidence of regimes are capable of 'capturing some of the essential features of the process involved in the formation of international governance systems'. Instead an 'institutional bargaining' approach is proposed which, although founded upon rationalistic assumptions, departs from neo-liberalism in certain important respects. These include a recognition of mixed motive bargaining by multiple actors under consensus rules which usually revolve around the negotiation of specific con-tested issues in a text. Most significantly, the conduct of bargaining within an institutional setting is much more conducive to agreement than in the kind of specific and self-contained negotiation that figures so largely in theories of bargaining behaviour. Emanating from Young's model are a series of hypo-theses about institutional bargaining which seem highly plausible with respect to the various global commons negotiations. These would include the pro-positions that the likelihood of success is increased by 'the availability of arrangements that all participants can accept as equitable', by the existence of 'clear cut and reliable compliance mechanisms' and by the stimulus provided by 'exogenous shocks' and 'entrepreneurial leadership' by individuals (Young, 1994: 106–115).

Institutional bargaining provides a much more proximate explanation of regime formation than that which has dominated much of this chapter. Looking across the spread of commons regimes a utilitarian hypothesis would appear to provide the best explanation for the incidence of well developed and effective institutions. There is clearly a relationship between mutual vulner-ability interdependence and effective regimes. In issue areas such as GSO, stratospheric ozone, maritime pollution and to a lesser extent Antarctica, the functional interests of users are served by institutions that allow orderly management of the commons. Contrary to Hobbesian assumptions and in line with findings related to small-scale commons, such institutions do not necessarily require imposition. They are stable and self-sustaining and rational users will see the sense in subscribing to them. Because of the costly disorder that would attend their collapse or decline they acquire value in themselves, particularly for those with a substantial stake in using the commons. This is something which, as demonstrated in the 'equity in orbit' campaign or in the London negotiations on the Montreal Protocol, can, under certain circumstances, be turned into a source of political leverage by the disadvantaged.

The contention that effective regimes do not require creation and supervision by an hegemonic actor does not mean that some members will not wield disproportionate influence or be able to benefit from a position of structural power. In most of the regimes under consideration there was fairly marked evidence of the way in which norms and rules emanating from US practices and legislation were translated to the international level.

Much of the discussion of regime creation and change has centred upon 'structural' explanation in the shape of the realist-inspired hegemonic stability thesis. There is little evidence to support the latter, except in the negative sense that the United States can wield what amounts to a veto. Depending upon the issues at stake, a number of other governments also enjoy such blocking potential. Most of the regimes under consideration have been extended and deepened over the last 20 years and some have displayed radical changes in direction. Yet these appear to have been almost entirely unrelated to any of the shifts in the overall power structure that have had such momentous consequences elsewhere. The relevant power relations are, thus, issue specific. Beyond making this point and observing that to be effective regime rules should normally, but not always, be congruent with the essential interests and power capabilities in the issue area, it is impossible to generalize.

As with Young's institutional bargaining model, discussed above, explaining the specifics of agenda setting and regime change in particular cases will require recognition of a plurality of interests and domestic, governmental and transnational players. The recent history of environmental regime building and change reveals that, contrary to the fixed motivational assumptions of both realism and liberal utilitarianism, shifting public values and their articulation by what might loosely be described as idealist political movements cannot be discounted. Much more attention has been paid to the rather elitist politics of epistemic communities and their role in the manipulation of the knowledge base of regimes. Cognitive change has been significant especially as it affects perceptions of mutual vulnerability, but so has value change and the two are not always coincident (the disputes over the future of whaling provide an illustration). There are compelling reasons to consider the role of epistemic communities but they should be tempered by the realization that such communities are one amongst a number of interests in a complex political process.

Some connections should be made between these rather specific explanations and generalized utilitarian or structural approaches. First, it is worth considering that while interdependence may provide the key to understanding the incidence of effective and stable regimes, perceptions of mutual vulnerability are neither fixed nor objective. They will be socially constructed and, in the cases under review, dependent upon a scientific understanding of the commons problem. Without an appreciation of the role of anthropogenic chemical emissions in the depletion of the stratospheric ozone layer, there was no issue area and no shared conception of common fate interdependence. In issue areas concerned with marine populations and pollution, it was a belated understanding of the finite nature of the commons that led to a realization of interdependence.

Second, the political processes of agenda setting and regime change occur within structural constraints. An appreciation of the structural position of the United States will, for example, direct the attention of both analyst and activist to public opinion and legislative and bureaucratic politics in that country. Thus epistemic community theorists realize that the engineering of cognitive change only has significance within a power structure. In the ozone case the really important achievement of the epistemic community was 'capturing the United States and DuPont' (Haas, 1992b: 224).

Notes

1. For a full-length study of the ITU which locates it as a core instrument of global hegemony see Lee (1996).
2. This establishes an ecosystem approach and requires the US to take steps to bring international arrangements into line with the legislation. The Act and the findings of the associated Marine Mammals Commission which it created have driven US policy towards the IWC, starting with repeated attempts to institute a moratorium during the 1970s. Its effect was extended out to 200 miles by the 1976 Fisheries Conservation and Management Act which also applies to the 1969 Endangered Species Act – banning the taking and importation of whales. A range of amendments to the legislation has provided a strengthened, if not always effective, legal basis for the imposition of sanctions against violators of IWC regulations. The legislation and its impact is discussed in Birnie (1985: 537–544). Australia and New Zealand adopted similar marine mammal protection legislation at the end of the 1970s.
3. It may be possible to exempt some issue areas from the necessity of hegemonic control by claiming that they do not exhibit the public goods problem. As argued in Chapter 1, common sink resources tend to exhibit this characteristic, but not others. For a full exposition of the theoretical underpinnings of the hegemonic stability thesis see Snidal (1985).
4. Interestingly the etymology of the word hegemony supports this interpretation. The original Greek 'hegemon' was a leader and the word 'hegemonia' is translated as 'authority'.
5. This is what Keohane (1984: 34) refers to as the 'basic force model' where, 'The theory of hegemonic stability predicts that the more one such power dominates the world political economy, the more cooperative will interstate relations be'. See also Porter & Brown (1991: 23).
6. Paterson (1996a: 94–101) considers the validity of the hegemonic stability thesis at some length in the climate change case and concludes that it is simply inapplicable.

7. Keohane & Nye (1977: 54–58) developed the international organization model of regime change whereby 'networks norms and institutions' were significant factors in their own right, which could explain why predictions based on overall power or issue structural models were not always fulfilled.
8. From the US Commerce Department's NTIA. See NTIA (1983) *Long Range Goals in International Telecommunications and Information: An Outline for US Policy*, US Senate, Committee on Commerce, Science and Transportation, 98th Congress.
9. The relevant cases are drawn from what Ostrom (1990: 211) calls the 'zero condition', in which the regime is relatively remote from an indifferent central government. This point about a perception of mutual harm is coupled to the proviso that most will be affected in similar ways by changes and have low discount rates for the future. Further down the list are 'relatively low information, transformation and enforcement costs'. As with international regimes, there is clearly some relationship between such institutional costs and the value placed upon the outcomes facilitated by the institution. Where the domestic requirements may differ from the international is in the existence of shared generalized norms of reciprocity which can be used as 'initial social capital' (although shared norms of behaviour are of course a crucial component of international regimes) – yet they are unlikely to be as deep and enduring in a relatively heterogeneous international system.
10. These findings about the 1993 decision derive from research by Lasse Ringius (1997).
11. This work has mainly been associated with Peter Haas, but the idea of epistemic communities is traceable to Ruggie (1975) and before that to Foucault's concept of an episteme.
12. It is argued that current writing on epistemic communities differs from neo-functionalism in that there is no intention to study the transfer of authority from national to supranational levels but rather to observe the impact of epistemic communities upon 'styles of policy-making and changes in reasoning' (Haas, 1992a: 12).
13. Johnston (1988: 205) further recognizes that 'To put ones trust in this kind of circuitry or networking is to return to "functionalism", or at least to the more modern ideas of the "neo-functionalists" in political science'.

9

Conclusion

This book started with the question of how the global commons could be effectively governed in the absence of world government. The realist answer is, by and large, that they cannot because of the structural anarchy of the interstate system. Under Hobbesian and individualistic assumptions about human behaviour commons 'tragedies' can only be avoided either by privatization or the imposition of rule by a government or some form of hegemonic substitute. Usually such arguments contrast the 'anarchic' condition of the international system with the supposed order of (notional) domestic systems under a sovereign government. However, arguments about the governance of domestic commons polarize in the same way. The followers of Hume and Hardin make the same assumptions as realists, while their opponents at domestic and international levels argue that self-managed governance institutions can exist and thrive in the absence of a government or a hegemon.

Regime analysis has provided a means of investigating the institutions of cooperation and governance that may exist at the global level alongside their, very much better understood and documented, counterparts at the local. The survey in Chapters 3 to 6 of this book demonstrates that, contrary to anarchic assumptions about the international system, a dense and interrelated network of governance institutions exists.

As we have seen in Chapter 7 there are serious problems and uncertainties surrounding their effectiveness. Indeed, a sceptic could make out the case that, in gross terms at least, an inverse relationship exists between the building of international institutions and the reduction of environmental degradation. Since 1972 some 130 new multilateral environmental agreements (MEAs) have been signed but the empirical evidence suggests that 'the forces that cause environmental degradation continue unabated' (Wapner, 1995: 47). Subsequent to the 1992 UNCED there was widespread disillusionment at the contrast between the scale of diplomatic activity and the paucity of real achievement. To add insult to injury, green activists perceived that the environmental movement itself had been highjacked by governments and the transnational corporate sector. As Sachs (1993: xv) wrote: 'ecology – understood as the philosophy of a social movement – is about to transform itself from a knowledge of opposition to a knowledge of domination'.

In response a significant academic critique has emerged which may be termed radical political ecology. It has various roots in deep green ecology, anarchism, the bioregional movement and Marxist political economy. Radical political ecologists argue that orthodox International Relations, which takes the anarchical states system as its problematic, is incapable of comprehending the global socio-economic phenomenon of environmental change. The whole enterprise of international cooperation and regime theorizing in particular is fundamentally flawed. It embodies a 'top down' managerialist approach based upon an atomistic science antithetical to ecological holism. It privileges the state in an era of globalization and actually obstructs action to nurture the commons. The actual site of degradation and real human cooperation is essentially local. Local commons arrangements are continually threatened with enclosure by the globalizing forces of capitalist development (*The Ecologist*, 1993). The equation between local and global commons is not only false but, when utilized by the governments and corporate representatives assembled at UNCED, pernicious. Cooperation between state governments is actually harmful because it necessarily reflects their collective interest in economic growth and a development paradigm which must ultimately result in environmental collapse. International environmental agreements are, at best, a means for governments to placate public opinion by 'doing something without doing anything' (Evans, 1998). Thus, for radical political ecologists operating within International Relations the well-worn debates between realists and neo-liberal-institutionalists about the possibility and circumstances of interstate cooperation are essentially pointless.[1]

The foregoing summary hardly does justice to the range and intensity of radical political economy.[2] But it does serve to illustrate how such thinking poses a fundamental challenge to the validity of international regimes for the sustainable management of the world environment and indeed to the very concept of global commons. Rather than simply ignore or deny the arguments of radical political ecology it is, in the context of the present study, important to engage with them in at least two respects. First, the assertion of the incompatibility between local and global commons prompts a closer analysis of whether the assumed isomorphic relationship[3] between local and global institutions, which has underpinned the argument advanced so far, is actually tenable. Second, the attack on the relevance of international cooperation ought to stimulate a reconsideration of the precise functions of global commons institutions in terms of what they can achieve in relation to other levels of governance. Put simply, this involves the actual, as opposed to theoretical, relationship between international, regional and local commons institutions.

Global and Local Commons

Although there are evident structural similarities between the commons at different spatial scales, and the arguments about cooperation are essentially the same, there are also six apparent differences which might serve to invalidate comparison.

First, it may be argued that local commons and their governance institutions represent a natural reality embedded in a particular ecology and economy – a bioregion. The close identity of human beings with such bioregions provides the key to their sustenance through fostering the institutional capacities of local communities (McGinnis, 1999: 2). An indicator of the natural quality of many local commons regimes as studied by Ostrom (1990) and others is their great longevity. Many have institutions which have survived for hundreds of years and which in one case of a Swiss pastoral commons are traceable back to 1224 (Ostrom, 1990: 62). By contrast, many global commons regimes appear as artificial constructs unnaturally cobbled together over a very brief period. But they also, it is argued, devalue and enclose the local, serving to 'maintain the webs of power that are currently stifling commons regimes':

> One cannot legislate the commons into existence; nor can the commons be reclaimed simply by adopting 'green techniques' such as organic agriculture, alternative energy strategies or better public transport – necessary and desirable though such strategies often are. Rather, commons regimes emerge through ordinary people's day-to-day resistance to enclosure, and through their efforts to regain mutual support, responsibility and trust that sustain the commons. (*The Ecologist*, 1993: 196–197)

The accusation that global commons regimes are recent and artificial constructs inevitably applies to recent attempts to grapple with the consequences of industrialization and technological change, as in the atmospheric regimes, but it is hardly a criticism. The focus on formal construction of new international institutions should not blind us to wider meaning of the regime concept or to the fact that some of the global regimes are of very long standing, even if they have had to adapt to rapid and recent technological change. The oceans regime developed through the accretion of understandings and practices just as any historic local regime might have done. A fascinating recent example of developing coordination of behaviour between users of a commons without, as yet, formally codified rules is provided by the orbital debris case.

Underlying the critique mounted by bioregionalists and *The Ecologist* is the seductive idea that there is something inherently real and natural about local-scale institutions that cannot be replicated at other scales. In fact, most human landscapes are inevitably unnatural in the sense that they are the product of human cultivation. One of the ironies of the debate about genetically modified organisms is that this most extreme form of human intervention in natural processes offers the possibility, through higher yields and reduced pesticide use, of restoring some landscapes to their primeval condition (Sagoff, 1991). The very idea of the commons is a social construct. As such it will take on a different meaning as economic, social and physical conditions change. That some local commons regimes may have somehow survived since time immemorial does not necessarily demonstrate that they are appropriate to contemporary conditions. More likely they represent an historical anachronism and a romantic desire to return to a pre-industrial world. The complex of technological, economic and cultural change encapsulated in the concept of globalization means that the local can never again be truly 'local' in the sense of isolation. The rise of NGO activity in relation to whaling or Antarctica also provides a potent demonstration that while people may be physically related to a particular bioregion they are more than capable of sustaining an intense identity with global phenomena. Utilization of the space commons has provided much of the electronic infrastructure allowing such identities to be constructed. A binary opposition which sets natural and virtuous local commons against attempts to construct global institutions seems increasingly futile under conditions of globalization. As Lipschutz (1999: 103) has argued, an understanding of the 'nesting' of commons institutions at various levels is far more appropriate:

> We can thus envision local systems of production and action – the immediate sources of environmental damage – as being nested within larger ones. These local resource regimes in turn, are part of economic, cultural and social relations of resource users and polluters rather than either being discrete or totally aggregated arrangements. Such user networks, moreover, are embedded in overlapping – but not necessarily coterminous – social, political, economic and physical spaces.

A second and related criticism of global commons institutions targets their state-centric character. 'Global' is in one sense a misnomer because they are, in reality, interstate institutions, attempting to superintend what are often very loosely described as global problems. The World Bank may call its environmental fund the Global Environmental Facility but it is actually an international fund run by an intergovernmental organization. UNCED may have been styled the Earth Summit and even hosted a 'global forum', but it was in reality a very large interstate diplomatic gathering at which non-state actors remained at the margin. The formal members of the various commons institutions that have been described are the 'state parties'. During the development of the climate change regime state governments made very sure that bodies such as the IPCC remained firmly 'intergovernmental'.

Despite various assaults on their competence nation states remain legally pre-eminent in the world system and it is therefore unsurprising that they should be the parties to global commons agreements. However, the argument is that they cannot be expected to rise above their own short-term national and electoral concerns. Ecology becomes the plaything of *realpolitik*:

> The rhetoric, which ornaments conferences and conventions, ritually calls for a new global ethic but the reality at the negotiating table suggests a different logic. There, for the most part, one sees diplomats engaged in a familiar game of accumulating advantages for their countries, eager to outmanoeuvre their opponents, shrewdly tailoring environmental concerns to interests dictated by their country's economic position. (Sachs, 1993: 12)

One might respond that if governmental representatives were not to do so they would be derelict in their duty. Regimes created by states will serve their interests, especially in relation to preservation of autonomy (Zacher with Sutton, 1996). However, the interests pursued are hardly monolithic and they have been subject to alteration sometimes by active public campaigning by environmentalists. The extent of such change in terms of the increased salience of environmental questions on governmental agendas has been chronicled by Brenton (1994) – a direct participant. There is also the suggestion, implicit in much critical commentary, that while those involved in local institutions or NGOs naturally represent their constituents or stakeholders, nation state governments are incapable of doing so. Unlike NGOs and informal local institutions, an increasing number of governments are in actual practice held to account by their electorates, who may choose not to privilege the sustenance of the commons above their immediate economic interests.

The discussion in Chapter 8 demonstrates how a realist analysis is simply inadequate. True, there are many instances of the pursuit of naked electoral or economic advantage, particularly in the emergence of the climate change regime. Yet, what is more remarkable is the way in which governmental perceptions have changed, and outcomes have occurred which were unthinkable in terms of stated national interests at the outset. In Brenton's (1994: 251) view, 'The whole history of international environmental action has been of arriving at destinations which looked impossibly distant at the moment of departure'.

At the same time there is a serious risk, perhaps exemplified by the deliberations of the United Nations Commission for Sustainable Development, that operating through intergovernmental organizations and Conferences of the Parties will become almost an end in itself. As has often been pointed out, the environmental costs of many such gatherings in terms of paper consumed and air travel are hardly commensurate with their, often purely declaratory, outcomes. This is only to argue that international cooperation in relation to the commons has a variety of functions for participants, involving status, organizational budgets, travel and even social intercourse. Some at least of these functions must be performed, admittedly at lesser cost, by local institutions as well.

A third line of attack concerns the presumed 'managerialism' of global commons regimes. Conventionally the global commons problematic has been set up in terms of the ability of a fragmented states system to *manage* global systems (Hurrell & Kingsbury, 1992: 1). For Saurin (1996: 79–80) this can never be the purpose of critical scholarship. It remains, however, the *raison d'être* of much of the academic work on regimes and their implementation that has been cited in the preceding chapters of this book. From a radical standpoint such work, which fails to investigate causation and the relationships between contemporary capitalism and the degradation of the commons, is essentially complicit in a managerial project sponsored by governments and corporations. Managerialism at the global level stands, once again, in sharp contrast to the voluntarism and interpersonal cooperation exhibited by local commons regimes (*The Ecologist*, 1993: 172–195). International institutions are pervaded by a 'top down' economistic approach to planetary problems which involve the kind of bureaucratic methodologies developed within nation states and where the objective is not to question how overconsumption in developed societies can be reigned in but how sustainable development can be achieved. In contrast to local regimes that may often be based on a 'vernacular knowledge' complementing local ecologies, global commons regimes are science driven. This is indisputable, for one of the key issues in all of the global regimes surveyed has been how to improve the generation of scientific knowledge and its direct application to the principles and rules of the regime.

The next assertion is that 'global management means global policing' (*The Ecologist*, 1993: 132). In terms of the existing global regimes this surely mistakes the hubristic rhetoric of some politicians, on the necessity to intervene to save the planet, for the sober day-to-day reality of international cooperation. Far from being able to enforce international-level decisions and alter behaviour at the local level, the analysis in Chapter 7 demonstrates the almost complete absence of such a capability. Absence of government and the kind of managerial 'command and control' potential that it is supposed to possess, provides the strongest parallel between international and local commons institutions.

A fourth contrast between the local and global commons highlights the close and symbiotic relationship between local ecologies and institutions. The domain of such regimes is appropriate. This kind of fit between institutions and problems is difficult to achieve at the global level both horizontally and vertically. The horizontal dimension refers to the ways in which international regimes only provide partial coverage of a commons problem and there are gaps between different and sometimes competing institutions. Regimes have been created at various times on the basis of sometimes narrowly constructed issue areas to leave, with the exception of the Antarctic case, something of a disjointed patchwork. Viewed from an ecological perspective this means that they can rarely be wholly effective because important sources of degradation and drivers of change lie outside their domain. A recurring theme has been the incongruity between fragmented issue areas and institutions and what Brundtland (WCED, 1987: 310) described as the 'real world of interlocked economic and ecological systems'. The ramifications of global climate change are so extensive, reaching to the core of modern industrial society, that it is difficult to see how effective international institutions can be erected and how they could be anything more than an adjunct to much broader change.

In this respect a gap of particular importance is that which exists between many of the environmental regimes that have been described and the international trade regime centred upon the GATT rules and WTO. There are some evident difficulties because the principles of the latter with its encouragement of free trade and complex rules, serving as a disarmament agreement amongst national trade protectionists, appear ignorant of and even antithetical to the preservation of the commons (von Moltke, 1997). A very significant academic and policy debate about the relationship between trade and the environment has been under way (Esty, 1994; Brack, 1998) and the WTO itself has created a Committee on Trade and the Environment. An immediate issue has been the need to resolve direct contradictions arising from 'GATT illegal' clauses in environmental agreements, notably the trade sanctions available under the Montreal Protocol (Brack, 1996). Also on the agenda is the non-recognition by the GATT of the legitimate right of governments to discriminate against products in terms of their environmentally damaging processes and production methods. At a more profound level, radical political ecologists have argued that there is an absolute incompatibility between the trade regime and the preservation of the global commons, simply because it fuels the engine of economic growth which is the main instrument of environmental degradation.

Improving the horizontal integration of the various commons regimes both amongst themselves and with cognate trade and monetary regimes has not proved to be an easy task as various attempts to simplify and coordinate the UN system have demonstrated. Institutions at both the local and global level tend to exhibit inertia and persistence and there is a danger that coordination attempts will simply introduce more organizations into an already over-crowded system. The evidence from the present study suggests that however desirable it may seem to negotiate commons issues on a comprehensive and inclusive basis, as in UNCLOS III, the political practicalities suggest that a more effective institution is likely to emerge from the kind of tight focus that characterized the creation of the ozone regime.

On the vertical dimension, it is clear that some global commons regimes are remote from the activities and behaviour they seek to regulate. This is not the case for the orbit/spectrum or space debris regimes, where spacecraft designers and operators are very much part of the institutions. It is the case for the marine pollution or climate change regimes. Here attempts are made, through intergovernmental agreement, to regulate myriad human activities occurring at the local scale. The question of the appropriate level of governance is taken up in the final part of this chapter, but it is worth noting that even here the global/local divide is not as clear cut as might initially be assumed. As Lipschutz (1999) argues, the pressures of economic and technological change bearing upon local commons institutions will tend to force them to adapt through a reconstruction of their domain. The focus of this transformation, which will bring in new participants and link interests together, will not be a specific resource *per se* but the changing political economy of its utilization.

A fifth apparent contrast between the local and the global concerns a significant barrier to collective action encountered by Hume's hypothetical neighbours in their attempts to drain a meadow. As the numbers involved increase, so do the difficulties of coordination and the 'transaction costs'. Theorists like Keohane (1984) have argued that it is a primary function of regimes to reduce such costs. This is clearly so, for it would be difficult to imagine how users could share the spectrum without the cost-effective guidance provided by the Radio Regulations. Yet there is also the simple question of scale, particularly when designing or changing a regime. Large numbers make coordination and institution building difficult. Most of the successful local commons regimes seem to have been relatively small. They also enjoy the benefits of community and a whole set of related common understandings and social obligations:

> Small scale communities are more likely to have the formal conditions required for successful and enduring collective management of the commons. Among these are the visibility of common resources and behaviour towards them; feedback on the effects of regulations; widespread understanding and acceptance of the rules and their rationales; the values expressed in these rules (that is equitable treatment of all and the protection of the environment); and the backing of values by socialization, standards and strict enforcement. (Ostrom, 1985 cited in *The Ecologist*, 1993: 16–17)

How can global institutions be in any way comparable? It is usual to think in state-centric terms, equating states with individuals and the problems of persuading over 180 governments, of widely different cultural and linguistic backgrounds, to pull together. Thus one might expect exclusive regimes, like that for Antarctica, to be more effective than those inviting universal membership. There is little apparent basis for such an assertion and the demand, in the UN General Assembly, for universal membership plus the need to involve all those having a stake in the commons makes such solutions unattainable and undesirable.

However, it may be that a concentration on states misses the point. Global regimes, like any other regimes, involve human social interaction. The real equivalent to the face-to-face communities of farmers and fishermen involved in local commons management are the equally restricted groups of specialists, the technical elites perhaps, who are engaged in the everyday operation of the various global regimes. The engineers on the consultative and technical committees of the ITU, the Antarctic Scientists on SCAR and their govern-mental and NGO counterparts provide examples. These transnational com-munities are usually fairly small. The humble Antarctic krill may constitute the largest biomass on earth, but a deep interest in its biology is something that unites a rather limited group of people mostly with some connection to the CCAMLR. The transnational orbital debris community comprises about 200 individuals world-wide.[4] These are real communities where participants inter-act regularly, and as such they are capable of fulfilling some of the functions ascribed to local communities in the operation of regimes. To provide one small example from the climate change regime, national greenhouse gas inventories under the FCCC are compiled by a restricted group of govern-mental specialists. They meet each other regularly, 'attend the same confer-ences, chat with each other and compete'. In the words of one participant it would, under these circumstances, be 'almost impossible to cheat and very difficult to fake your inventory'.[5] These communities might qualify for the label 'epistemic' except that they tend to be comprised of governmental officials and scientific personnel. They will, in any event, tend to have good connections with the wider epistemic community in their field. If this analysis is correct, investigating the sociology of such 'global regime communities', and the way in which they facilitate the development and application of rules, may prove to be as significant as the study of epistemic communities.

A sixth and final point of contention between local and global regimes is that the former, to function well, are claimed to operate on equitable prin-ciples. The concept of equitable shares of a common resource or equitable rights of access is evidently important in legitimizing commons institutions under conditions of 'self-organization'. Whether this is always the case, given the feudal roots of many local institutions, is a moot point. As was observed in the introductory discussion, equity considerations tend to be absent from treatments of the 'tragedy of the commons' which assume the mutual inter-dependence of users and neglect the very real 'tragedy of dispossession'. It is, however, equally questionable whether equity rules for the distribution of a resource necessarily promote sustainability for there will always be pressure to solve distributional problems by increasing overall exploitation rather than reducing individual shares.

The scale of inequality witnessed at world level can hardly have any counterpart in local communities, particularly as these are unlikely to span the gap between affluent 'centres' and disadvantaged 'peripheries' that exists within many countries. The politics of North–South confrontation over devel-opment issues, responsibility for environmental degradation and a fair alloca-tion of shares to global resources has been ever-present in the cases surveyed in this book. Its particular manifestation has been the campaign to institute the Common Heritage of Mankind (CHM) as a cardinal regime principle. Actual achievement in this respect has been limited to a Moon Treaty, unratified by the spacefaring nations, a much weakened set of arrangements for the seabed and the limited access to orbit agreement of the ITU of 1988.

There are at least two reasons why achievement in this area has been so minimal. First of all, it was associated with the NIEO and 'socialistic' centralizing approaches to economic management. With the end of the Cold War and the disintegration of the Soviet Union and its extensive international alliances, such ideas ceased to have appeal. The extent of the dominance achieved by liberal 'free market' ideas is reflected by a clause in the 1994 UN General Assembly Resolution relating to the seabed common heritage provisions of the Third LoS Convention, which recognizes:

> that political and economic changes, including in particular a growing reliance on market principles, have necessitated the re-evaluation of some aspects of the regime for the Area and its resources.[6]

Second, there was always a contradiction between the idea of 'permanent sovereignty over resources' and CHM. The principle of 'permanent sovereignty' represented a key demand by newly independent Southern governments resolved to ensure that their national economic assets were not expropriated by the industries of the developed world. As we have seen, this demand was adroitly combined with the corollary of environmental responsibility in Principle 21 of the 1972 Stockholm Declaration. By the 1992 Rio Summit, however, those who had advocated CHM in respect of the seabed Area were pressing for 'permanent sovereignty' with respect to biodiversity, forests and climate change on the grounds that CHM would 'impair their proprietary rights over resources under their jurisdiction'. 'The outcome of this process was that the concept of "common concern" rather than "common heritage" was accepted' (Schrijver, 1997: 389).

However, this hardly exhausts the significance of North–South inequality issues in global commons regimes. The interdependence of regime participants has proved, on a number of occasions, to be a potent source of leverage in obtaining outcomes favoured by the G77 which, as was argued in Chapter 8, could not be attributed to the structural position of developing countries or their relational power resources. It is equally clear that no effective climate change regime can be created without addressing the question, not only of compensatory payments in terms of technology transfer and 'capacity building', but also, in the longer term, of the assumption of emissions reduction commitments by Southern economies. Given the historic responsibilities of the OECD countries for climate change and their continuing high consumption levels, this will be a hard bargain both to strike and to justify in terms of any notion of equity.

Multi-layer Governance

Two contradictory arguments have obscured discussion of the global commons. On the one hand, there is the easy assertion that global problems necessarily require global solutions in terms of international institutions. On the other, there is a radical rejection of all forms of international cooperation in favour of 'grass roots' action by local regimes. Both were probably promoted by the Rio process, the former by the hyperbole of the Earth Summit, the latter by the inevitable disillusionment that attended its aftermath. The point that is obscured is not the relative virtues of global or local-scale regimes, as discussed above, but the critical relationship between them.

Taken to its logical extreme, the idea that the scale of the institution must match the scale of the problem leads to what Wapner (1995) has described as 'supra statism' or the demand to centralize authority in a world government. This appears to have been the solution to global commons tragedies advocated by Hardin himself (Wapner, 1995: 54–55). Most analysts, however, tend to agree that a move to world government is both unlikely and undesirable and it has been the argument of this book that the existing interstate system can develop governance regimes for the commons. It is, however, not the case that every problem which has a global dimension is best addressed by a full-scale international regime. The tendency in the UN system has been to call for an international conference on virtually every issue. While such meetings may serve a purpose in terms of placing and keeping certain items on both national and international agendas, they are likely to be counterproductive if they lead to failed attempts at governance or become mere 'talking shops' (Brenton, 1994: 266). Even the largest developed world governments are already overloaded by the sheer volume of international environmental business and their shortage of 'in house' expertise has created significant opportunities for NGOs to exert influence.

The other argument, often associated with radical political ecology, is that contemporary states are either so tainted or ineffective as to render enlightened commons governance at the international level impossible. Ekins (1993), for example, argues that there is a necessary contradiction between the state with its 'scientism' and 'developmentalism' and 'the people'. The 'artificial' and 'extremely unresponsive' nation state has been involved in 'a global replay of the English Enclosure Acts of the seventeenth century which deprived the common people of their means of livelihood' (Ekins, 1993: 206). The state is intimately associated with capitalism's modern project that drives the destruction of the commons. International cooperation must, therefore, reflect this. It follows that preservation of the commons can only be achieved through struggle between 'grassroots movements for global change' and state authorities.

Although there is much force in the radical critique, it is one thing to argue that the scale of global change requires a complete reversal of the economic growth trajectory of capitalism allied to a total overhaul of the international political system, and quite another to contemplate it as practical politics. The state, although challenged by the forces of globalization and in many instances devoid of the most rudimentary capabilities, has not yet withered away. Were it to be further weakened, it is more than questionable whether 'grass roots' movements would be able to fashion instruments to restrain the degradation wrought by corporations or straightforward criminality.

The cases in this book provide some grounds for optimism that state authorities are capable of responding to popular and environmental concerns – at least on occasion. One of the virtues of conducting regime analysis is that it forces the observer to think systematically about principles and norms. This reveals the extent of change across the global commons regimes. The campaign for common heritage may have met with little success, but other shifts at the level of norms and principles have been striking. Alterations in the whaling and Antarctic regimes have been of sufficient magnitude to be describable in terms of thoroughgoing regime change. The marine pollution regime has been transformed to outlaw activities that were once permitted. There have been changes which cannot be wholly accounted for in terms of either the influence of corporate capitalism or the pursuit of narrowly defined national interest. In short, the nation state is not always and necessarily antagonistic to effective action to safeguard the commons.

The difficulties in evaluating the effectiveness of regimes are very real, but an attempt was made in Chapter 7 to provide a rough ranking of the cases in terms of various criteria. This yielded a set of effective implementation regimes that were capable of governing the global commons even if there were continuing problems with their sometimes narrow domain. They share certain characteristics. When governments are involved directly in using or regulating the use of the global commons then the regimes that they create can approximate the functions of local commons institutions. Notable here are the common property resource regimes (CPRs) created to manage what were 'open access' *res nullius* resources: the orbit/spectrum regime, the Antarctic regime, and belatedly the seabed regime. For the 'common sinks' there have been achievements in the provision of international public goods in terms of a reduction of polluting behaviour at sea (if not overall maritime pollution) and the beginnings of the restoration of the stratospheric ozone layer. UNEP's judgement as to the importance of international cooperation in coping with the depletion of the stratospheric ozone layer is worth quotation:

> Without the Protocol, by the year 2050 ozone depletion would have risen to at least 50% in the northern hemisphere's mid latitudes and 70% in the southern mid latitudes, about 10 times worse than current levels. The result would have been a doubling of UV-B radiation reaching the earth in northern mid latitudes and a quadrupling in the south. The amount of ozone-depleting chemicals in the atmosphere would have been five times greater. The implications of this

would have been horrendous: 19 million more cases of non-melanoma cancer, 1.5 million cases of melanoma cancer, and 130 million more cases of eye cataracts (http://www.unep.org/ozone, 1999).

The ozone case along with the orbit/spectrum, Antarctic and marine pollution case are also notable for the increasing sophistication and complexity of their rule-making activity and their ability to adapt to shifting scientific knowledge. The space debris regime appears to be moving along the same path. Similarly, there is much in the institutionalist argument that regimes, once created, have an educative function which can subtly alter the ways in which participants perceive their interests.

Above all there is one characteristic shared by all the effective regimes and which differentiates them from issue areas where regimes either failed to develop at all or where regime arrangements were weak or purely declaratory. This is simply a shared perception of mutual vulnerability interdependence amongst the major participants. Interdependence may involve 'keeping the system going' and facilitating commerce and communications in areas beyond the reach of sovereignty as in the classical high seas regime and that for orbit and spectrum. Pollution and scarcity provide new forms of 'common fate interdependence' which also serve to underpin effective regimes.

These are issue areas where it is difficult to envisage any other regulatory system not involving cooperation at the interstate level. Only state governments have the requisite incentives, authority and resources to control transboundary commerce, to provide global public goods and to police the oceans and outer space. For the moment, at least, state authorities also maintain their right to authorize space launches and the usage of the frequency spectrum. They, rather than the private sector, also fund the research upon which (despite challenges to 'scientism') our developing understanding of the biophysical and socio-economic characteristics of the global commons depend. There is an essentially similar point to be made about the GEF funding and 'capacity building' resources.

However, even amongst the effective intergovernmental regimes it would be highly misleading to restrict analysis to states and the mutual interdependence of their national interests. Another characteristic shared by most of the successful regimes, and accounting in large measure for significant changes in their norms and principles, is their penetration by the agents of what has come to be known as 'global civil society'. The best known are the environmentally active NGOs. There is ample evidence from the cases that they have 'linked the global with the local' in the ways outlined by Princen & Finger (1994). They have been directly responsible for the politicization of the environment at national and transnational levels, which has exerted pressure on governments to revise national policy. The changes in the whaling and Antarctic regimes are inexplicable in any other way and NGO 'agenda setting' activity has been significant in the ozone, marine pollution and climate regimes. They have been less visible in the space regimes. Here other non-state actors from the corporate sector have tended to dominate. Neither should it be forgotten that it is the collective interest of the developed world chemical industries which, as much as public concern for the effects of ozone depletion, holds the Montreal Protocol regime in place. NGOs also serve to substitute for governments in areas where they lack capacity. This involves the provision of expertise and advice but also, as in the marine pollution, Antarctic and whaling cases, the monitoring of compliance.

The problem cases occur where there is a recognition of interdependence and strong demand for action but where governments are remote from the myriad causes of degradation. Climate change and marine pollution, in the broadest sense, fall into this category. The first edition of this book concluded that the scale and complexity of such problems, and especially climate change, was probably beyond the capabilities of the international political system as presently constituted. Regimes, the institutions which serve to coordinate the activities of governments in the absence of centralized authority, it was asserted, might simply not be enough. The intergovernmental arrangements described here probably constitute only one component of a fully effective commons regime to cope with a problem such as climate change. Such a perspective would recognize that there are only some problems and functions for which international cooperation provides an appropriate solution. Instead intergovernmental regimes constitute one part of a multi-layered governance system.

In the case of the emerging regime for climate change the FCCC and the Kyoto Protocol may be seen as the 'peak regime' (Lipschutz, 1999) at the apex of a pyramid of governance reaching right down to very localized institutions. The functions appropriate to the international level are familiar from the cases in this book, the coordination and review of national emissions reduction commitments and related policies and measures, the provision of scientific knowledge and compensatory funding. In future the development of the Kyoto 'mechanisms' will also invest the regime with a trade regulating function. Neither should the significance of the FCCC in generating principles and norms be neglected. In line with the well worn slogan of 'think globally, act locally' it is important to conceptualize and legitimize policies at the highest level. The other necessary role of an international regime is to manage the problems of horizontal integration. In this case better coordination not only with the stratospheric ozone regime, but also with the trade and monetary regimes, is needed in order to set up patterns of incentive for behaviour which supports rather than opposes the mitigation of climate change.

We already have evidence that some apparently global environmental problems are best dealt with at a regional rather than a global level. In the case of the FCCC the European Union has demonstrated how a regional burden-sharing arrangement under its 'bubble' of differentiated emissions commitments can include countries at different levels of economic development in a common target for greenhouse gas reductions.

While the FCCC, regional bodies and governments can set targets and alter patterns of incentive, it is at the local level that climate change inducing behaviour actually occurs. It is also at this level that the strongest commons management institutions can be created with high levels of identification with local ecologies and popular participation. There are already 'numerous examples of subnational policy and program development aimed at mitigating climate change' (Feldman & Wilt, 1999: 146).

International Relations scholars have, over the last 25 years, been engaged in the study of regimes at the international level. For the global commons such regimes have, on occasion, proved to be institutionally dynamic and effective. They perform functions that, in the current world system, cannot be performed at any other level. As this book has attempted to demonstrate, they are functionally comparable to local commons regimes, but they cannot be a substitute for them. Neither is the reverse true. The critical question for the future is not the superiority of global or local institutions but the fit between them, for both must play a role in the sustenance of the commons.

Notes

1. In the UK during the 1990s the main point of contention between those who study the International Relations of the environment has been precisely this – the relevance of international cooperation. Elsewhere, and particularly in the United States, the debate has been between neo-realist and liberal accounts of cooperation – essentially a subset of a wider controversy within International Political Economy. However, there has been an emerging challenge to the mainstream from those like Wapner (1996) who have questioned the state-centric bases of conventional scholarship and explored the concept of world civic society and politics.
2. For a good introduction to this extensive literature the reader is referred to Paterson (1996b).
3. An isomorphic relationship is one in which two or more entities have the same structure and whose corresponding parts have similar properties and relations. Thus it is argued that, although existing at vastly different scales, local and global commons regimes share certain essential institutional features.
4. Normal attendance at IADC meetings is 100, while the transnational community rises to 400 if all researchers in the field are added (interview, Richard Tremayne Smith, British National Space Centre (BNSC), January 1999).
5. Participant at ESRC Global Environmental Change Programme research dissemination meeting on the FCCC, London, 5 September 1997.
6. UN General Assembly, A/RES/48/263, 17 August 1994.

References

Andresen, S., 1998, 'The Making and Implementation of Whaling Policies: Does Participation Make a Difference?', in Victor *et al.* (eds), pp. 431–476.

Andresen, S. & Østreng, W., 1989, *International Resource Management: The Role of Science and Politics*, London, Belhaven.

Axelrod, R., 1984, *The Evolution of Cooperation*, New York, Basic Books.

Baker, D., 1989, 'Remote Sensing: A Political Football in the Sky', *Intermedia*, 17, 1, pp. 19–26 and 17, 2, pp. 48–54.

Barkin, J.S. & Shambaugh, C.E., 1996, 'Common Pool Resources and International Environmental Politics', *Environmental Politics*, 5, 3, Autumn, pp. 429–447.

Barnes, J.N., 1991, 'Protection of the Environment in Antarctica: Are Present Regimes Enough?', in Jørgensen-Dahl & Østreng (eds), pp. 186–228.

Barston, R.P., 1980, 'The Law of the Sea Conference: The Search for New Regimes', in Barston & Birnie (eds), pp. 154–168.

Barston, R.P. & Birnie, P. (eds), 1980, *The Maritime Dimension*, London, George Allen & Unwin.

Beck, P.J., 1991, 'The Antarctic Resource Conventions Implemented: Consequences for the Sovereignty Issue', in Jørgensen-Dahl and Østreng (eds), pp. 229–265.

Beck, P.J., 1998, 'The United Nations and Antarctica, 1996: maintaining consensus towards the millennium', *Polar Record*, 34, 188, pp. 39–44.

Beck, P.J. & Dodds, K., 1998, *Why Study Antarctica?*, CEDAR Research Paper no. 26, Egham, Surrey, Royal Holloway College.

Beeby, C.D., 1991, 'The Antarctic Treaty System: Goals, Performance and Impact', in Jørgensen-Dahl & Østreng (eds), pp. 4–24.

Bellany, I., 1987, 'Controlling Military Activity in Space', in Kirby & Robson (eds), pp. 223–238.

Benedick, R.E., 1991, *Ozone Diplomacy: New Directions in Safeguarding the Planet*, Cambridge, MA, Harvard University Press.

Berkes, F. (ed.), 1989, *Common Property Resources Ecology and Community-Based Sustainable Development*, London, Belhaven.

Bernauer, T., 1995, 'The Effect of International Environmental Institutions: How We Might Learn More', *International Organization*, 49, 2, Spring, pp. 351–377.

Birnie, P.W., 1980a, 'Contemporary Maritime Legal Problems', in Barston & Birnie (eds), pp. 168–189.

Birnie, P.W., 1980b, 'The Law of the Sea Before and After UNCLOS I and UNCLOS II', in Barston & Birnie (eds), pp. 8–26.

Birnie, P.W., 1985, *International Regulation of Whaling: From Conservation of Whaling to Conservation of Whales and Regulation of Whale Watching*, New York, Oceana Publications Inc., Vols I–II.

Birnie, P.W. & Boyle, A.E., 1992, *International Law and the Environment*, Oxford, Clarendon Press.

Boehmer-Christiansen, S., 1993, 'Science Policy, The IPCC and the Climate Convention: The Codification of a Global Research Agenda', *Energy and Environment*, 4, 4, pp. 362–407.

Boehmer-Christiansen, S., 1996, 'The International Research Enterprise and Global Environmental Change: Climate-Change Policy as a Research Process', in Vogler & Imber (eds), pp. 171–195.

Brack, D., 1996, *International Trade and the Montreal Protocol*, London, RIIA/ Earthscan.

Brack, D. (ed.), 1998, *Trade and Environment: Conflict or Compatibility*, London, RIIA/Earthscan.

Brack, D. & Grubb, M., 1996, *Climate Change: A Summary of the Second Assessment Report of the IPCC*, London, RIIA Briefing Paper No. 32, July.

Branscomb, A.W., 1979, 'Waves of the Future: Making WARC Work', *Foreign Policy*, 34, pp. 139–148.

Brenton, T., 1994, *The Greening of Machiavelli: The Evolution of International Environmental Politics*, London, RIIA/Earthscan.

Bretherton, C. & Vogler, J., 1999, *The EU as a Global Actor*, London, Routledge.

Brierly, J.L., 1963, *The Law of Nations: An Introduction to the Law of Peace*, Oxford, Clarendon Press.

Brown, A.D., 1991, 'The Design of CRAMRA: How Appropriate for the Protection of the Environment?', in Jørgensen-Dahl & Østreng (eds), pp. 110–119.

Brown, C.V. & Jackson, P.M., 1990, *Public Sector Economics*, 4th edition, Oxford, Basil Blackwell.

Brown, S. & Fabian, L.L., 1975, 'Towards Mutual Accountability in the Non-Terrestrial Realms', *International Organization*, 29, pp. 877–892.

Brown, S. *et al.*, 1977, *Regimes for the Ocean, Outer Space and Weather*, Washington, Brookings.

Buck, S.J., 1989, 'Multi-Jurisdictional Resources: Testing a Typology for Problem Structuring', in Berkes (ed.), pp. 127–163.

Burrows, W.E., 1986, *Deep Black: Space Espionage and National Security*, New York, Random House.

Burton, J.W., 1972, *World Society*, Cambridge, Cambridge University Press.

Caldwell, L.K., 1984, *International Environmental Policy Emergence and Dimensions*, 2nd edition 1990, Durham, NC, Duke University Press.

Caldwell, L.K., 1990, *Between Two Worlds: Science, The Environmental Movement and Policy Choice*, Cambridge, Cambridge University Press.

Cameron, J., Werksman, J. & Roderick, P., (eds), 1996, *Improving Compliance with International Environmental Law*, London, Earthscan.

Campbell, J.L., Rogers Hollingsworth, J. & Lindberg, L.N. (eds), 1991, *Governance of the American Economy*, Cambridge, Cambridge University Press.

Carroll, J.E. (ed.), 1988, *International Environmental Diplomacy*, Cambridge, Cambridge University Press.

Chayes, A. & Chayes, A.H., 1991, 'Adjustment and Compliance Processes in International Regulatory Regimes', in Mathews (ed.), pp. 280–309.

Chayes, A. & Chayes, A.H., 1996, *The New Sovereignty*, Cambridge, MA, Harvard University Press.

Chenard, S., 1990, 'Space: Dirty and Dangerous?', *Interavia Space Markets*, 6, 5, pp. 253–256.

Christol, I.C.Q., 1982, *The Modern International Law of Outer Space*, Oxford, Pergamon.

Codding, G.A., 1981, 'Influence in International Conferences', *International Organization*, 35, 4, pp. 715–724.

Codding, G.A. & Landa-Fournais, F., 1991, 'Policy Options: Reorganization and Reform', in Mohan & Vogler (eds), pp. 43–48.

Codding, G.A. & Rutkowski, A.M., 1982, *The International Telecommunication Union in a Changing World*, Dedham, MA, Artech House Inc.

Commission on Global Governance, 1995, *Our Global Neighbourhood*, Oxford, Oxford University Press.

Darman, R.E., 1978, 'The Law of the Sea: Rethinking U.S. Interests', *Foreign Affairs*, 56, pp. 373–395.

Cox, R.W., 1983, 'Social Forces, States and World Orders: Beyond International Relations Theory', reprinted in Kratochwil, F. & Mansfield, E. (eds) 1994, *International Organization: A Reader*, New York, Harper Collins, pp. 343–363.

Davis, P.B., 1998, 'Understanding Visitor Use in Antarctica: The Need for Site Criteria', *Polar Record*, 34, 188, pp. 45–52.

Demac, D. *et al.*, 1985, *Equity in Orbit: The 1985 ITU Space WARC: A Background Paper*, London, IIC.

Demac, D. *et al.*, 1986, *Access to Orbit: After the 1985 ITU Space WARC: A Follow-Up Report*, London, IIC.

Denman, D.R., 1984, *Markets Under the Sea?*, London, IEA/Hobart Paperback 17.

Donnelly, J., 1986, 'International Human Rights: A Regime Analysis', *International Organization*, 40, 3, Summer, pp. 599–642.

Downs, G.W., Rocke, D.M. & Barsoom, P.N., 1996, 'Is the Good News About Compliance Good News About Cooperation?', *International Organization*, 50, 3, Summer, pp. 379–406.

Drewry, D.J., 1987, 'The Antarctic Physical Environment', in Triggs (ed.), pp. 6–27.

Driver, P.A., 1980, 'International Fisheries', in Barston and Birnie (eds), pp. 27–53.

Eckersley, R., 1992, *Environmentalism and Political Theory: Toward an Ecocentric Approach*, London, University College London Press.

The Ecologist, 1993, *Whose Common Future? Reclaiming the Commons*, London, Earthscan.

Edwards, D., 1988, 'Review of the Status of Implementations and Development of Regional Arrangements on Co-operation in Combating Marine Pollution', in Carroll (ed.), pp. 229–274.

Ekins, P., 1993, *A New World Order: Grassroots Movements for Global Change*, London, Routledge.

Elliott, L., 1994, *International Environmental Politics: Protecting the Antarctic*, London, Macmillan.

Esty, D., 1994, *Greening the GATT: Trade Environment and the Future*, Washington, Institute for International Economics.

Evans, A., 1998, 'Doing Something without Doing Anything: International Environmental Law', in Jewell, T. and Steele, L. (eds), *Law in Environmental Decision-Making*, Oxford, Clarendon Press, pp. 207–227.

Falk, R.A., 1971, *This Endangered Planet*, New York, Random House.

Falk, R.A., 1991, 'The Antarctic Treaty System: Are There Viable Alternatives?', in Jørgensen-Dahl & Østreng (eds), pp. 399–414.

Fawcett, J.E.S., 1984, *Outer Space: The New Challenge to Law and Policy*, Oxford, Clarendon Press.

Feldman, D.L. & Witt, C.A., 1999, 'Climate-Change Policy from a Bioregional Perspective: Reconciling Spatial Scale with Human and Ecological Impact', in McGinnis (ed.), pp. 133–154.

Ferreira, V., 1993, 'The United Nations Framework Convention on Climate Change: What Next?', in Poole & Guthrie (eds), pp. 239–246.

Gallagher, B. (ed.), 1989, *Never Beyond Reach: The World of Mobile Satellite Communications*, London, Inmarsat.

Gehring, T., 1990, 'International Environmental Regimes; Dynamic Sectoral Legal Systems', *Yearbook of International Environmental Law*, 1, pp. 35–36.

Gibbs, C.J.N. & Bromley, D.W., 1989, 'Institutional Arrangements for Management of Rural Resources: Common Property Regimes', in Berkes (ed.), pp. 22–32.

Goerz, G. & Diehl, P.F., 1992, 'Toward a Theory of International Norms: Some Conceptual and Measurement Issues', *Journal of Conflict Resolution*, 36, 4, December, pp. 634–664.

Gorove, K. & Kamenetskaya, E., 1995, 'Tensions in the Development of the Law of Outer Space', reprinted in Ku & Diehl (eds), 1998, pp. 473–506.

Gray, C.B. & Rivkin, D.B., 1991, 'A "No Regrets" Environmental Policy', *Foreign Policy*, 83, pp. 47–65.

Gray, C.S., 1983, 'Space is Not a Sanctuary', *Survival*, 15, 5, pp. 194–204.

Greene, O., 1993a, 'International Environmental Regimes: Verification and Implementation Review', *Environmental Politics*, 2, 4, pp. 156–173.

Greene, O., 1993b, 'Limiting Ozone Depletion: The 1992 Review Process and the Development of the Montreal Protocol', in Poole & Guthrie (eds), pp. 269–280.

Greene, O., 1996, 'Environmental regimes: effectiveness and implementation review', in Vogler & Imber (eds), pp. 196–214.

Greene, O., 1998, 'The system of implementation review in the ozone regime', in Victor et al. (eds), pp. 89–136.

Grubb, M., 1990, 'What to do About Global Warming: The Greenhouse Effect Negotiating Targets', *International Affairs*, 66, 1, pp. 67–90.

Grubb, M., 1992, *The Greenhouse Effect: Negotiating Targets*, London, RIIA.

Grubb, M. & Anderson, D. (eds), 1995, *The Emerging International Regime for Climate Change Structures and Options After Berlin*, London, RIIA.

Grubb, M. & Steen, N. (eds), 1991, *Pledge and Review Processes: Possible Components of a Climate Convention*, London, RIIA.

Grubb, M. et al., 1993, *The Earth Summit Agreements: A Guide and Assessment*, London, RIIA/Earthscan.

Gulland, J.A., 1987, 'The Antarctic Treaty System as a Resource Management Mechanism', in Triggs (ed.), pp. 116–129.

Haas, E.B., 1964, *Beyond the Nation State: Functionalism and International Organization*, Stanford, CA, Stanford University Press.

Haas, E.B., 1980, 'Why Collaborate? Issue Linkage and International Regimes', *World Politics*, XXXII, 3, pp. 357–405.

Haas, E.B., 1990, *When Knowledge is Power: Three Models of Change in International Organization*, Berkeley, University of California Press.

Haas, E.B., Williams, M.P. & Babai, D., 1977, *Scientists and World Order: The Uses of Technological Knowledge in International Organizations*, Berkeley and Los Angeles, University of California Press.

Haas, P.M., 1990a, 'Obtaining International Environmental Protection Through Epistemic Communities', *Millennium*, 19, 3, pp. 347–364.

Haas, P.M., 1990b, *Saving the Mediterranean: The Politics of International Environmental Cooperation*, New York, Columbia University Press.

Haas, P.M., 1992a, 'Introduction: Epistemic Communities and International Policy Coordination', *International Organization*, 46, 1, pp. 1–36.

Haas, P.M., 1992b, 'Banning Chlorofluorocarbons: Epistemic Community Efforts to Protect Stratospheric Ozone', *International Organization*, 46, 1, pp. 187–224.

Haas, P.M., Keohane, R.O. & Levy, M. (eds), 1993, *Institutions for the Earth: Sources of Effective International Environmental Protection*, Cambridge, MA, MIT Press.

Haggard, S. & Simmons, B.A., 1987, 'Theories of International Regimes', *International Organization*, 41, 3, pp. 491–517.

Hamelink, C.J., 1994, *The Politics of World Communication*, London, Sage.

Hardin, G., 1968, 'The Tragedy of The Commons', reprinted in Hardin, G. & Baden, J. (eds), 1977, *Managing the Commons*, San Francisco, W.H. Freeman & Co., pp. 16–30.

Hardin, G., 1978, 'Political Requirements for Preserving our Common Heritage', in H.P. Bokaw (ed.), *Wildlife and America*, Washington, DC, Council on Environmental Quality, pp. 310–317.

Haron, M., 1991, 'The Ability of the Antarctic Treaty System to Adapt to External Challenges', in Jørgensen-Dahl & Østreng (eds), pp. 299–308.

Heap, J., 1988, 'The Role of Scientific Advice for the Decision-Making Process in the Antarctic Treaty System', in Wolfrum, R. (ed.), *Antarctic Challenge III*, pp. 21–28.

Henkin, L., 1979, *How Nations Behave: Law and Foreign Policy*, 2nd edition, New York, Columbia University Press.

Herr, R.A., 1996, 'The Changing Roles of Non-Governmental Organisations in the Antarctic Treaty System', in Stokke & Vidas (eds), pp. 91–110.

Hills, J., 1989, 'The Internationalization of Domestic Telecommunications Law and Regulations: US Industrial Policy on a World Wide Basis', in Garnham, N. (ed.), *European Telecommunications Policy Research*, Amsterdam, IOS, pp. 46–57.

Holdgate, N.W., 1987, 'Regulated Development and Conservation of Antarctic Resources', in Triggs (ed.), pp. 117–127.

Hollick, A.L., 1981, *US Foreign Policy and The Law of The Sea*, Princeton, NJ, Princeton University Press.

Hopgood, S., 1998, *American Foreign Environmental Policy and the Power of the State*, Oxford, Oxford University Press.

Houghton, J.T. *et al.* (eds), 1990, *Climate Change, The IPCC Scientific Assessment: Report Prepared for IPCC by Working Group I*, Cambridge, Cambridge University Press.

Houghton, J.T. *et al.* (eds), 1996, *Climate Change 1995: The Science of Climate Change*, Cambridge, Cambridge University Press.

Hudson, H., 1985, 'Access to Information Resources: The Development Context of the Space WARC', *Telecommunications Policy*, March, pp. 23–30.

Hulme, M. *et al.*, 1992, *Climate Change Due to the Greenhouse Effect and its Implications for China*, London, WWF.

Hume, D., 1740, *A Treatise of Human Nature*, Selby Bigge, L.A. (ed.), 1888, Oxford, Clarendon Press.

Hurrell, A. & Kingsbury, B. (eds), 1992, *The International Politics of the Environment*, Oxford, Clarendon Press.

IEA (Institute of Economic Affairs), 1994, *Global Warming: Apocalyse or Hot Air?*, London, IEA.

Imber, M., 1988, 'International Institutions and The Common Heritage of Mankind:

Sea, Space and Polar Regions', in Taylor, P. & Groom, A.J.R. (eds), *International Institutions at Work*, London, Pinter, pp. 150–166.

Imber, M., 1993, 'Too Many Cooks? The Post-Rio Reform of the UN', *International Affairs*, 69, 1, pp. 55–70.

Jacobson, H.K., 1974, 'ITU a Pot-Pourri of Bureaucrats and Industrialists', in Cox, R.W. & Jacobson, H.W. (eds), *The Anatomy of Influence: Decision Making in International Organizations*, New Haven and London, Yale University Press, pp. 58–101.

Jäger, J. & Ferguson, H.L. (eds), 1991, *Climate Change: Science, Impacts and Policy: Proceedings of the Second World Climate Conference*, Cambridge, Cambridge University Press.

Jakhu, R.S., 1983, 'The Evolution of the ITU's Regulatory Regime Governing Space Radiocommunication Services and the Geostationary Satellite Orbit', *Annals of Air and Space Law*, 8, pp. 381–407.

Jasani, B. & Lee, C., 1984, *Countdown to Space War*, London, Taylor & Francis, SIPRI.

Johnston, D.M., 1988, 'Marine Pollution Agreements: Successes and Problems', in Carroll (ed.), pp. 199–206.

Jones, R.J.B. & Willetts, P. (eds), 1984, *Interdependence on Trial: Studies in the Theory of Reality of Contemporary Interdependence*, London, Frances Pinter.

Jønnson, B., 1972, 'The UN Institutional Response to Stockholm: A Case Study in the International Politics of Institutional Change', *International Organization*, 20, 2, pp. 255–301.

Jønsson, C., 1987, *International Aviation and the Politics of Regime Change*, London, Frances Pinter.

Jordan, A., 1998, 'The Ozone Endgame: The Implementation of the Montreal Protocol in the United Kingdom', *Environmental Politics*, 7, 4, Winter, pp. 23–52.

Jørgensen-Dahl, A. & Østreng, W. (eds), 1991, *The Antarctic Treaty System in World Politics*, Basingstoke, Macmillan.

Joyner, C., 1990, 'Antarctic Treaty Diplomacy: Problems, Prospects and Policy Implications', in Newson, D. (ed.), *The Diplomatic Record 1989/1990*, Washington, DC, Georgetown University.

Joyner, C., 1992, *Antarctica and the Law of the Sea*, Nijhoff, Den Haag.

Katz, J.E. (ed.), 1985, *People in Space*, New Brunswick, NJ, Transaction Inc.

Kay, D.A. & Jacobson, H.K. (eds), 1983, *Environmental Protection: The International Dimension*, New Jersey, Allanheld Osmun & Co.

Keohane, R.O., 1984, *After Hegemony: Cooperation and Discord in the World Political Economy*, Princeton, NJ, Princeton University Press.

Keohane, R.O. & Nye, J.S. (eds), 1973, *Transnational Relations and World Politics*, Cambridge, MA, Harvard University Press.

Keohane, R.O. & Nye, J.S., 1977, *Power and Interdependence: World Politics in Transition*, Boston, Little, Brown.

Kirby, S. & Robson, G. (eds), 1987, *The Militarisation of Space*, Sussex, Wheatsheaf.

Krasner, S.D. (ed.), 1983, *International Regimes*, Ithaca, NY, Cornell University Press.

Krasner, S.D., 1985, *Structural Conflict: The Third World Against Global Liberalism*, Berkeley, University of California Press.

Kratochwil, F.V., 1989, *Rules, Norms and Decisions: On the Conditions of Practical and Legal Reasoning in International Relations and Domestic Affairs*, Cambridge, Cambridge University Press.

Ku, C. & Diehl, P.F. (eds), 1998, *International Law: Classic and Contemporary Readings*, Boulder, CO, Lynne Rienner.

Larminie, G., 1991, 'The Mineral Potential of Antarctica: The State of the Art', in Jørgensen-Dahl & Østreng (eds), pp. 79–93.

Larschan, B. & Brennan, B.C., 1983, 'The Common Heritage of Mankind, Principle in International Law', *Columbia Journal of Transnational Law*, 21, pp. 305–337.

Laursen, F., 1982, 'Security Versus Access to Resources; Explaining a Decade of US Ocean Policy', *World Politics*, XXXIV, 2, pp. 197–205.

Laver, M., 1984, 'The Politics of Inner Space: Tragedies of Three Commons', *European Journal of Political Research*, 12, pp. 59–71.

Laver, M., 1986, 'Public, Private and Common in Outer Space: Res Extra Commercium or Res Communis Humanitatis, Beyond the High Frontier', *Political Studies*, XXXIV, 3, pp. 359–373.

Laws, R.M., 1987, 'Scientific Opportunities in the Antarctic', in Triggs, G.D. (ed.), pp. 28–50.

Lee, C., 1987, *War in Space*, London, Sphere.

Lee, K., 1996, *Global Telecommunications Regulation: A Political Economy Perspective*, London, Pinter.

Lipschutz, R.D., 1999, 'Bioregionalism, civil society and global environmental governance', in McGinnis (ed.), pp. 101–120.

Lipson, C., 1991, 'Why Are Some International Agreements Informal?', *International Organization*, 45, 4, pp. 495–538.

Litfin, K., 1994, *Ozone Discourses: Science and Politics in Global Environmental Cooperation*, New York, Columbia University Press.

Lundberg, O., 1993, 'Telecommunications for Development: Mobile Satcoms', *Intermedia*, 21, 6, pp. 12–13.

Lyall, F., 1989, *Law and Space Telecommunication*, Dartmouth and Aldershot, Gower.

McGinnis, M.V. (ed.), 1999, *Bioregionalism*, London, Routledge.

McGraw, D. (ed.), 1991, *International Law and Politics*, Philadelphia, University of Pennsylvania Press.

McPhail, T.L., 1981, *Electronic Colonialism: The Future of International Broadcasting and Communication*, London, Sage.

Magdelenat, J.L., 1985, 'The Controversy Over Remote Sensing', in Katz (ed.), pp. 129–139.

Maitland, D., 1984, The *Missing Link: Report of the Independent Commissions for World-Wide Telecommunication Development*, Geneva, ITU.

Mansbach, R.W. et al., 1976, *The Web of World Politics: Non State Actions in the Global System*, London, Prentice Hall.

Mathews, J.T. (ed.), 1991, *Preserving the Global Environment: The Challenge of Shared Leadership*, New York, W.W. Norton.

Meadows, D.H. et al., 1992, *Beyond the Limits: Global Collapse or Sustainable Development?*, London, Earthscan.

Messer, K. & Breth, R., 1991, 'Towards Firmer Institutionalization of the ATS? The Role of the Consultative Meeting and the Issue of a Permanent Secretariat', in Jørgensen-Dahl & Østreng, pp. 379–398.

Middleton, N., O'Keefe, P. & Moyo, S., 1993, *Tears of the Crocodile: From Rio to Reality in the Developing World*, London, Pluto Press.

Miller, J.D., 1981, *The World of States*, London, Croom Helm.

Milner, H., 1992, 'International Theories of Cooperation Among Nations: Strengths and Weaknesses', *World Politics*, 44, 3, pp. 466–496.

Mitchell, R.B., 1993, 'International Oil Pollution of the Oceans', in Haas et al. (eds), pp. 184–247.

Mitchell, R.B., 1994, *Intentional Oil Pollution at Sea: Environmental Policy and Treaty Compliance*, Cambridge, MA., MIT Press.
Mitchell, R.B., 1996, 'Compliance Theory: An Overview', in Cameron *et al.* (eds), pp. 3–28.
Mitrany, D., 1975, *The Functional Theory of Politics*, London, LSE/Martin Robertson.
Mohan, C.V. & Vogler, J. (eds), 1991, *Sharing Spectrum in the Digital Age – WARC '92*, London, International Institute of Communication and Friedrich-Ebert Stifting.
Morphet, S., 1996, 'NGOs and the Environment', in Willetts (ed.), pp. 116–146.
Morse, E.L., 1976, *Modernisation and the Transformation of International Relations*, New York, Free Press.
Morse, E.L., 1977, 'Managing International Commons', *Journal of International Affairs*, 31, 1, pp. 1–21.
Moss, N., 1991, *The Politics of Global Warming*, London Defence Studies, No. 9, London, Brasseys for the Centre for Defence Studies.
Naraine, M., 1985, 'Warc-Orb-85, Guaranteeing Access to the Geostationary Orbit', *Telecommunications Policy*, pp. 77–108.
Newell, P., 1998, 'Who "CoPed" Out in Kyoto? An Assessment of the Third Conference of the Parties to the Framework Convention on Climate Change', *Environmental Politics*, 7, 2, Summer, pp. 153–159.
Nitze, W., 1990, *The Greenhouse Effect: Formulating a Convention*, London, RIIA.
Nollkaemper, A., 1992, 'On the Effectiveness of International Rules', *Acta Politica*, 27, pp. 49–70.
North, D.C., 1990, *Institutions, Institutional Change and Economic Performance*, Cambridge, Cambridge University Press.
Nurmi, S., 1988, 'Issues and Problems in the Protection of the Marine Environment', in Carroll (ed.), pp. 207–227.
Ogley, R., 1984, *Internationalizing the Seabed*, Aldershot, Gower.
Olson, M., 1965, *The Logic of Collective Action: Public Goods and the Theory of Groups*, Cambridge, MA, Harvard University Press.
Ophuls, W., 1977, *Ecology and the Politics of Security: Prologue to a Political Theory of the Steady State*, San Francisco, W.H. Freeman.
Ostrom, E., 1990, *Governing the Commons: The Evolution of Institutions for Collective Action*, Cambridge, Cambridge University Press.
OTA (Office of Technology Assessment), 1986, *Strategic Defenses*, Princeton, NJ, Princeton University Press.
Oxman, B.H., Caron, D. & Buderi, C. (eds), 1983, *Law of the Sea*, San Francisco, Institute for Contemporary Studies.
Oye, K. (ed.), 1986, *Cooperation Under Anarchy*, Princeton, NJ, Princeton University Press.
Pardo, A., 1984, 'A New Order for the Oceans', *South*, London, May, pp. 73–74.
Parson, E.A., 1993, 'Protecting the Ozone Layer', in Haas *et al.* (eds), pp. 27–74.
Paterson, M., 1996a, *Global Warming and Global Politics*, London, Routledge.
Paterson, M., 1996b, 'Green Politics', in Burchill, S., Linklater, A. *et al.*, *Theories of International Relations*, London, Macmillan, pp. 252–274.
Pearce, D.W., 1991, *Blueprint 2: Greening the Global Economy*, London, Earthscan.
Pearce, D.W., Markandy, A.A. & Barbier, E., 1989, *Blueprint for a Green Economy*, London, Earthscan.
Pearce, D.W. *et al.*, 1991, *Blueprint 3: Measuring Sustainable Development*, London, Earthscan.
Peterson, M.J., 1992, 'Whalers, Cetologists, Environmentalists and the International Management of Whaling', *International Organization*, 46, 1, pp. 147–186.

Peterson, M.J., 1993, 'International Fisheries Management', in Haas *et al.* (eds), pp. 249–308.

Pirages, D., 1978, *Global Ecopolitics: The New Context for International Relations*, N. Scituate, MA, Duxbury.

Ploman, E.W., 1984, *Space Earth and Communication*, London, Pinter.

Poole, J.B. & Guthrie, R. (eds), 1993, *Verification 1993*, London, VERTIC, Brasseys.

Porter, G. & Brown, J.W., 1991, *Global Environmental Politics*, Boulder, CO, Westview Press.

Prescott, V., 1980, 'The Deep Seabed', in Barston & Birnie (eds), pp. 54–75.

Princen, T. & Finger, M., 1994, *Environmental NGOs in World Politics: Linking the Local and the Global*, London, Routledge.

Puissochet, J.P., 1991, 'CCAMLR – A Critical Assessment', in Jørgensen-Dahl & Østreng (eds), pp. 70–78.

Ramakrishna, K., 1990, 'North–South Issues, Common Heritage of Mankind and Global Change', *Millenium*, 19, 3, pp. 429–446.

Richardson, E.L., 1980, 'Power, Mobility and the Law of the Sea', *Foreign Affairs*, 58, pp. 902–919.

Ringius, L., 1997, 'Environmental NGOs and regime change: the case of ocean dumping of radioactive wastes', *European Journal of International Relations*, 3, 1, pp. 61–104.

Rittberger, V. (ed.), 1993, *Regime Theory and International Relations*, Oxford, Clarendon Press.

Roach, C., 1990, 'The Movement for a New World Information and Communication Order: A Second Wave?', *Media Culture and Society*, 12, 3, pp. 283–308.

Rose, G. & Rowland, E., 1993, 'Verification and Enforcement Under the International Convention for the Regulation of Whaling', in Poole & Guthrie (eds), pp. 259–269.

Rosenau, J.N. & Cziempiel, E.O. (eds), 1992, *Governance Without Government: Order and Change in World Politics*, Cambridge, Cambridge University Press.

Ruggie, J.G., 1975, 'International Responses to Technology: Concepts and Trends', *International Organization*, 29, 3, pp. 557–583.

Sachs, W. (ed.), 1993, *Global Ecology: A New Arena of Political Conflict*, London, Zed Books.

Sagoff, M., 1991, 'On Making Nature Safe for Biotechnology', in Ginzburg, L.R. (ed.), *Assessing Ecological Risks for Biotechnology*, Boston, Butterworth/Heinemann, pp. 341–365.

Sand, P.H., 1991, 'International Cooperation: The Environmental Experience', in Mathews (ed.), pp. 236–279.

Sand, P.H. (ed.), 1992, *The Effectiveness of International Environmental Agreements: A Survey of Existing Legal Instruments*, Cambridge, UNCED/Grotius.

Sands, P. (ed.), 1993, *The Greening of International Law*, London, Earthscan.

Sanger, C., 1986, *Ordering the Oceans: The Making of the Law of the Sea*, London, Zed Books.

Saurin, J., 1996, 'International Relations, Social Ecology and the Globalisation of Environmental Change', in Vogler & Imber (eds), pp. 77–98.

Sauvant, K.P. & Hasenflug, H. (eds), 1977, *The New International Economic Order*, London, Wilton House Press.

Schauer, W.H., 1977, 'Outer Space: The Boundless Commons?', *Journal of International Affairs*, 31, 1, pp. 67–80.

Schmidt, M.G., 1989, *Common Heritage or Common Burden?*, Oxford, Clarendon Press.

Schreuers, M.A. & Economy, E.C. (eds), 1997, *The Internationalization of Environmental Protection*, Cambridge, Cambridge University Press.

Schrijver, N., 1997, *Sovereignty Over Natural Resources: Balancing Rights and Duties*, Cambridge, Cambridge University Press.

Sebenius, J.K., 1984, *Negotiating The Law of The Sea*, Cambridge, MA, Harvard University Press.

Sebenius, J.K., 1993, 'The Law of The Sea Conference: Lessons for Negotiations to Control Global Warming', in Sjöstedt (ed.), pp. 189–216.

Simon, H., 1993, 'STAR TV and the Asian Television Revolution', *Intermedia*, Jan–Feb, 21, 1, pp. 6–7.

Sjöstedt, G. (ed.), 1993, *International Environmental Negotiations*, London, IIASA/Sage.

Skolnikoff, E.B., 1990 'The Policy Gridlock on Global Warming', *Foreign Policy*, 79 Summer, pp. 77–93.

Skolnikoff, E.B., 1993, *The Elusive Transformation: Science, Technology and the Evolution of International Politics*, Princeton, NJ, Princeton University Press.

Smith, A., 1980, *The Geopolitics of Information: How Western Culture Dominates the World*, London, Faber.

Smith, S., 1993, 'The Environment on the Periphery of International Relations: An Explanation', *Environmental Politics*, 2, 4, pp. 28–45.

Snidal, D., 1985, 'The Limits of Hegemonic Stability Theory', *International Organization*, 39, 4, Autumn, pp. 579–614.

Soroos, M.S., 1982, 'The Commons in The Sky: The Radio Spectrum and Geosynchronous Orbit as Issues in Global Policy', *International Organization*, 36, 3, pp. 665–677.

Soroos, M.S., 1986, *Beyond Sovereignty: The Challenge of Global Policy*, Columbia, University of South Carolina Press.

Soroos, M.S., 1991, 'The Evolution of Global Regulation of Atmospheric Pollution', *Policy Studies Journal*, 19, 2, pp. 115–125.

Soroos, M.S., 1997, *The Endangered Atmosphere: Preserving a Global Commons*, Columbia, University of South Carolina Press.

Sprinz, D. & Vahtorana, T., 1994, 'The interest based explanation of international environmental policy', *International Organization*, 48, 1, Winter, pp. 77–106.

Staple, G., 1986, 'The New World Satellite Order: A Report from Geneva', *American Journal of International Law*, 80, pp. 699–720.

Staple, G. et al., 1989, Reforming the Global Network: The 1989 ITU Plenipotentiary Conference, London, International Institute for Communications.

Stares, P.B., 1987, *Space and National Security*, Washington, DC, Brookings.

Stoett, P.J., 1997, *The International Politics of Whaling*, Vancouver, University of British Columbia Press.

Stokke, O.S., 1996, 'The effectiveness of CCAMLR', in Stokke & Vidas (eds), pp. 120–173.

Stokke, O.S., 1997, 'Regimes as Governance Systems', in Young (ed.), pp. 27–64.

Stokke, O.S., 1998, 'Nuclear Dumping in the Arctic Seas: Russian Implementation of the London Convention', in Victor et al. (eds), pp. 475–518.

Stokke, O.S. & Vidas, D. (eds), 1996, *Governing the Antarctic: The Effectiveness and Legitimacy of the Antarctic Treaty System*, Cambridge, Cambridge University Press.

Stone, C., 1993, 'Defending the Global Commons', in Sands (ed.), pp. 35–49.

Stowe, R.F., 1983, 'The Legal and Political Considerations of the 1985 World Administrative Radio Conference', *Journal of Space Law*, 11, 1 and 2, pp. 61–65.

Strange, S., 1983, 'Cave! hic dragones; A Critique of Regime Analysis', in Krasner (ed.), pp. 337–354.

Strange, S., 1987, 'The Persistent Myth of Lost Hegemony', *International Organization*, 41, 4, pp. 551–574.

Strange, S., 1988, *States and Markets: An Introduction to International Political Economy*, London, Pinter.

Suter, K., 1991, *Antarctica: Private Property or Public Heritage?*, London, Zed.

Taylor, M., 1976, *Anarchy and Cooperation*, London, John Wiley & Sons.

Taylor, M. & Ward, H., 1982, 'Chicken, Whales and Lumpy Goods', *Political Studies*, XXX, pp. 350–70.

Thomas, C., 1992, *The Environment in International Relations*, London, RIIA.

Tolba, M.K. & El-Kholy, O.A. *et al.* (eds), 1992, *The World Environment 1972–1992: Two Decades of Challenge*, London, UNEP/Chapman & Hall.

Tooze, R., 1990, 'Regimes and International Cooperation', in Groom, A.J.R. & Taylor, P. (eds), *Frameworks for International Cooperation*, London, Pinter, pp. 201–216.

Triggs, G.D. (ed.), 1987, *The Antarctic Treaty Regime: Law, Environment and Resources*, Cambridge, Cambridge University Press.

Underdal, A., 1994, 'Measuring and Explaining Regime Effectiveness', in Hveem, H. (ed.), *Complex Cooperation: Institutions and Processes in International Resource Management*, Oslo, Scandinavian University Press.

Underdal, A., 1995, 'The Study of International Regimes', *Journal of Peace Research*, 32, 1, pp. 113–119.

van der Lugt, C., 1997, 'An International Regime for the Antarctic: Critical Investigations', *Polar Record*, 33, 186, pp. 223–238.

Veljanovski, C., 1987, British Cable and Satellite Television Policies', National Westminster Bank Quarterly Review, November, pp. 28–40.

Victor, D.G., 1998, 'The Operation and Effectiveness of the Montreal Protocol's Non-Compliance Procedure', in Victor *et al.* (eds), pp. 137–176.

Victor, D.G., Raustiala, K. & Skolnikoff, E.B. (eds), 1998, *The Implementation and Effectiveness of International Environmental Commitments: Theory and Practice*, Cambridge, MA, IIASA/MIT Press.

Vicuna, F.O., 1991, 'The Effectiveness of the Decision-making Machinery of the CCAMLR: An Assessment', in Jørgensen-Dahl & Østreng (eds), pp. 25–42.

Vicuna, F.O., 1996, 'The effectivenes of the Protocol on Environmental Protection to the Antarctic Treaty', in Stokke & Vidas (eds), pp. 174–202.

Vidas, D., 1996, 'The Antarctic Treaty System in the International Community: An Overview', in Stokke & Vidas (eds), pp. 35–60.

Vogler, J., 1984, 'Interdependence, Power and the World Administrative Radio Conference', in Jones & Willetts (eds), pp. 200–225.

Vogler, J., 1992, 'The Global Commons: Space, Atmosphere and Oceans' in McGrew, A. & Lewis, P. *et al.* (eds), *Global Politics*, Oxford, Polity, pp. 118–137.

Vogler, J., 1996, 'The Structures of Global Politics', in Bretherton, C. & Ponton, G. (eds), *Global Politics: An Introduction*, London, Blackwell.

Vogler, J., 1999, 'The EU as an Actor in International Environmental Politics', *Environmental Politics*, 8, 3, Autumn, pp. 24–48.

Vogler, J. & Imber, M. (eds), 1996, *The Environment and International Relations*, London, Routledge.

von Moltke, K., 1997, 'Institutional Interactions: The Structure of Regimes for Trade and the Environment', in Young (ed.), pp. 247–272.

Waltz, K.N., 1979, *Theory of International Politics*, Reading, MA, Addison-Wesley.

Wapner, P., 1995, 'The State and Environmental Challenges: A Critical Exploration of Alternatives to the State System', *Environmental Politics*, 4, 1, Spring, pp. 44–69.

Wapner, P., 1996, *Environmental Activism and World Civic Politics*, Albany, NY, State University of New York Press.

Wapner, P., 1997, 'Governance in Global Society', in Young (ed.), pp. 65–84.

Ward, B. & Dubos, D., 1972, *Only One Earth: The Care and Maintenance of a Small Planet*, Harmondsworth, Penguin/André Deutsch.

Watts, A., 1992, *International Law and the Antarctic Treaty System*, Cambridge, Grotius.

Weale, A., 1992, *The New Politics of Pollution*, Manchester, Manchester University Press.

Webb, M.C. & Krasner, S.D., 1989, 'Hegemonic Stability Theory: An Empirical Assessment', *Review of International Studies*, 15, 2, pp. 183–98.

Werksmann, J., 1996, 'Designing a Compliance System for the UN Framework Convention on Climate Change', in Cameron *et al.* (eds), pp. 85–112.

Wijkman, P.M., 1982, 'Managing the Global Commons', *International Organization*, 36, 3, pp. 511–536.

Wilkinson, J. *et al.*, 1998, 'Orbital Debris: A Risk to Satellite Communications', *Intermedia*, 26, 1, March, pp. 20–22.

Wilkinson, J. *et al.*, 1999, 'Space Debris Mitigation: Reducing the Risk of Future Satellite Collision', *Intermedia*, 27, 2, May, pp. 38–43.

Willetts, P. (ed.), 1996, *Conscience of the World: The Influence of Non Governmental Organisation in the UN System*, London, Christopher Hurst for the David Davies Institute.

Williams, S.M., 1981, 'International Law Before and After the Moon Agreement', *International Relations*, 7, pp. 1168–1193.

Williams, S.M., 1985, 'Direct Broadcast Satellites and International Law', *International Relations*, 8, 3, pp. 245–269.

Woolf, L.S, 1916, *International Government*, London, Allen & Unwin.

World Bank, 1992, *World Development Report 1992: Development and the Environment*, New York, Oxford University Press.

WCED (World Commission on Environment and Development, The Brundtland Report), 1987, *Our Common Future*, Oxford, Oxford University Press.

Young, O.R., 1977, *Resource Management at the International Level*, New York, Nichols.

Young, O.R., 1982, *Resource Regimes: Natural Resources and Social Institutions*, Berkeley, University of California Press.

Young, O.R., 1983, 'Regime Dynamics: The Rise and Fall of International Regimes', in Krasner (ed.), pp. 93–114.

Young, O.R., 1989, *International Cooperation: Building Regimes for Natural Resources and the Environment*, Ithaca, NY, Cornell University Press.

Young, O., 1994, *International Governance: Protecting the Environment in a Stateless Society*, Ithaca, NY, Cornell University Press.

Young, O. (ed.), 1997, *Global Governance: Drawing Insights from the Environmental Experience*, Cambridge, MA, MIT Press.

Young, O.R. & Oroshenko, G. (eds), 1993, *Polar Politics: Creating International Environmental Regimes*, Ithaca and London, Cornell University Press.

Zacher, M.W. with Sutton, B.A., 1996, *Governing Global Networks: International Regimes for Transportation and Communication*, Cambridge, Cambridge University Press.

Zain-Azraai, A., 1987, 'Antarctica: The Claims of "Expertise" Versus "Interest"', in Triggs (ed.), pp. 211–217.

Index

a priori or *posteriori* rights vesting 31, 116–19
activity criterion *see under* Antarctic Treaty
Ad Hoc Group on the Berlin Mandate (AGBM) *see under* Framework Convention on Climate Change
aerosol propellants, US ban on CFCs as 125, 187
Agenda 21 34
  Antarctica absent from 80
  and marine pollution 58
  see also United Nations Conference on Environment and Development
agenda setting 29, 185, 202, 224–5
airspace, national 122
Andromeda Strain 105
Antarctic
  Convergence 73–5
  science community 204
  tourism 76
Antarctic and Southern Ocean Coalition (ASOC) 84
Antarctic Treaty, 1959 7
  activity criterion 81, 198
  CCAMLR 77–8, 82–5, 86, 89–90, 167, 178, 198
  CRAMRA 77–8, 82–5, 87–8, 90, 159, 168, 185
  demilitarization 200
  Madrid Protocol to 85, 88–90, 159, 163, 165, 168
  Recommendations 85, 89
  regime assessment 180

SCAR 90–1
  and territorial claims 73, 79
Antarctic Treaty Consultative Parties (ATCP), meetings 34, 83–5, 87, 158
Antarctic Treaty System (ATS) 79–80, 158, 163–5, 168, 185, 190–1, 198
AOSIS 26, 137
Area, the 47, 62–5
  see also deep seabed *and* Law of the Sea Convention
Argentina, and Antarctica 80
arms control 102–5, 123, 167
ASAT (Anti-satellite weapons) 102–4, 191
Asiasat 99
atmospheric testing 123
Atochem Corporation 27, 126
Australia, rejection of CRAMRA 87–8

Baghdad, satellite coverage of attack on 96
Bahia Paraiso, wreck of 89
'balance of power' 190
BBC World TV 111
behaviour modification 166–74
bioregions 214, 215
blue whale unit 53, 184
'Boat Paper' 67
Bogota Declaration 99
Brazil 192
Bretton Woods 185
British Antarctic Survey, and 'ozone hole' 126

Brundtland, Gro Harlem
  Commission 1, 15
  and whaling 51, 175
  see also sustainable development
bureaucratic politics 203-4
Bush, President George 137, 203
Byrd Resolution 195

C band *see* geostationary orbit
C^3I 95-6
Canada
  and LoS 47
  and radioactive debris from Kosmos
    954 105
carbon dioxide 122, 134-5
Carter administration 198
cetaceans 49
  see also whaling
CFCs (chlorofluorocarbons) 125-6, 131,
  187, 192, 199
Chile, and Antarctica 80
China
  carbon dioxide emissions 135
  and DBS 111
  and Montreal Protocol 127, 192,
    194
  space programme 97-9
'claims registries' 168
Clarke, Arthur C. 111
climate change 133-7, 199
  contested 176
  regime creation 181
climate modelling 134
cognitive
  change 204-7
  theories 40
CNN 96, 110
Cold War 101-2, 191
collective action problem 11-12, 175,
  197
  see also Hume, Olson, public goods,
    transaction costs
collision cascading 106
'common heritage of mankind' 158,
  175-9, 193, 221
  and atmosphere 122
  attempts to apply to Antarctica 80-2
  principles 31-2
  seabed as 6-7, 62-4
  see also Pardo

common pool resources 3-4
common property resource (CPR)
  4-5
  deep seabed as 63-4
  fisheries as 48-9
  frequency spectrum as 113
  regimes 13, 23, 223
common sinks 3, 16, 223
  atmosphere as 123
  oceans as greatest 55
commons
  defined 2
  enclosure 12, 17, 196
  English commons 2, 12, 13-15
  high seas 3
  local and global 3, 173, 175
  open access 4-5
  outer space 99
  property rules 3
  tragedy of 10-15
  see also global commons
compliance 168-74
  in Antarctic 89-90
  and climate regime 142-3
  and deep seabed 67
  function of regimes 38
  and marine pollution regimes 60-1
  and orbit/spectrum regime 117
  and ozone regime 132-3
  and whaling 54-5
consumer boycotts 193
Convention on Trade in Endangered
    Species (CITES), and whaling 52
cooperation 190, 196
COPUOS (UN Committee on the
    Peaceful Uses of Outer Space)
    100-1, 105, 108
cultural homogeneity 166

data provision 167, 169
DBS (direct broadcast by satellite) 95,
  109-11, 197
deep seabed 47, 61-9, 190, 198
  mining 61-2
  mining code 68
  Pioneer Investors 68
  regime assessment 180
  Seabed Council and Assembly 68
  Seabed Disputes Chamber 65
  see also Part XI

discount rate 198
dispute settlement 161
distribution, function of regimes 37
domestic politics, significance for regime
   creation and change 202–3
Donnelly, Jack 162–6
Du Pont Corporation 126, 199, 210

electromagnetic spectrum 7–8
enclosure 12, 15, 222
   of the high seas 45, 47
enforcement 38, 171–4 *see also under*
   compliance
enforcement regimes 163–5
Enterprise, the 66, 68–9
entry into force (EIF) 159
environmental impact assessment, in
   Antarctic 90
environmentalism 201–2
   and whaling 55
EOS (Earth Observation System)
   109
epistemic communities 28, 204–7, 220
   see also Haas, Peter
equity in orbit 114–19, 194, 200
European Community *see* European
   Union
European Space Agency 107
European Union 26
   and compliance 169, 171
   emissions 'bubble' 140–1, 150, 171,
      225–6
   and FCCC 137, 138, 140–2, 171,
      203
   and LoS 47
   and Montreal Protocol 126, 130
Exclusive Economic Zones (EEZs) 16,
   47, 193, 200
expansion bands 117

fisheries
   Antarctic 75 *see also under* Antarctic
      Treaty, CCAMLR
   collapse of stocks 48–9
   commissions 44, 48
flexibility, in regimes 160
Forests Convention, failure to negotiate
   193
four freedoms 44

'framework convention adjustable
   protocol' model 160
Framework Convention on Climate
   Change (FCCC) 136–44, 159, 160,
   163–5, 169, 171, 175–6, 188, 195,
   199, 225
   Ad Hoc Group on the Berlin Mandate
      (AGBM) 137, 140
   compensatory finance 141–2
   Conference of the Parties 139–40
   emissions trading 141
   flexibilty mechanisms 140–1
   Kyoto Protocol 124, 134, 139, 140,
      141, 171, 190, 195, 225
   narrowness of 25, 178–9
   national communications 142
   QUELROs 140
   SBSTA 143
frequencies, radio 112–13 *see also*
   electromagnetic spectrum *and* orbit/
   spectrum resource
functionalism 40, 153, 204

geostationary orbit (GSO)
   debris in 106
   defined 8
   efficient use of 176
   military use of X band 200
   planning of 111–17, 187, 193
   scarcity in 112, 176
Germany, and LoS 47
GESAMP 61
global, concept of 9
global atmospheric problems 124,
   146–7
global cellphones 96, 105
global commons 9
   defined 2, 6–7
global environmental change 1, 9
   and Antarctica 76–7
Global Environmental Facility (GEF)
   142, 224
global warming 133–4
   projected effects 14–15
governance 153, 162, 196
   defined 17
   multi-layered 222, 225
green politics 201
greenhouse effect, enhanced 8,
   133–6

greenhouse gases 134-5
Greenpeace International 28
  in Antarctica 89, 171
  and radioactive waste dumping 61
  and whaling 50
Grotius 44, 184
Group of 7 (G7) 137
Group of 77 (G77) 26, 45-6, 193-5,
  221
  and deep seabed 63
  and UNCLOS III 46
Gulf War 95-6, 104

Haas, Peter 40, 204-7
halons 125
Hardin, Garret 10-12, 212, 222
  critique of 12-14
  see also commons, tragedy of
HCFCs (hydrochlorofluorocarbons) 125,
  129, 134, 146
hegemonic
  leadership 189
  stability 39, 188-9, 196
hegemony 187-90
  limits of 188
'high politics' 200
high seas 44-7
Hobbes, Thomas 188, 200, 212
  idiom of rational choice 10, 39
  Leviathan 12, 188
holism 178
human rights regimes 162, 166
Hume, David, and collective action 10,
  12, 212, 219

icebergs 76
ICI Corporation 126
idealism 153, 200
implementation regimes 163-5, 184
incrementalism, in regime change 41
India
  and climate change 135
  and GSO 114, 200
  and Montreal Protocol 127, 129,
    192
information
  and deep seabed 66
  function of regimes 38
  and whaling 54-5

information flow, failure of regime
  building 108-11, 179
Inmarsat 99, 101, 113
Intelsat 99, 101, 113
Inter-Agency Orbital Debris
  Coordination Committee (IADC)
  107, 108, 171
interdependence 11, 184, 192, 193-5,
  199, 201, 209
  as common fate 11, 14, 197, 198,
    224
  as mutual vulnerability 11, 197, 198-9,
    224
intergenerational equity 134, 201
Intergovernmental Panel on Climate
  Change (IPCC) 28, 134-6, 143-4,
  205-7, 215
'intermestic' relationships 204
International Atomic Energy Agency
  167
International Convention for the
  Regulation of Whaling (ICRW) *see*
  under whaling
International Council for the Exploration
  of the Seas (ICES) 59
International Council of Scientific Unions
  (ICSU) 76, 90, 136
International Frequency Registration
  Board (IFRB) 116, 117, 167
International Geophysical Year 1957/58
  76, 79
international law 155-61, 167, 172
  concept of regime 21
  customary status of EEZs 47
  and International Relations 161
International Maritime Organization, and
  marine pollution 59
International Negotiating Committee
  (INC) 136-7
International Observation System (for
  whaling) 54
international orders 23
international organization 35-6
International Seabed Authority 65, 161,
  165
  Preparatory Commission 67-8
international scientific cooperation
  alphabet soup of 144
  and cognitive explanations 204-7
  and science as a special interest
    205

International Telecommunication Union
(ITU) 156, 159–60, 162, 163, 168–9
development of 111–14
engineering subculture 205–6
organization and reforms 115
and private sector 27
rule-making and standard setting
115–19
International Whaling Commission
(IWC) *see under* whaling
Iraq, denied space intelligence in
1990–91 104
Iridium *see* Motorola Corporation
issue area 23–5
Antarctica as 78
and climate change 134–6, 192
and ecological holism 24–5, 178
maritime 46–8
space 101–2
issue-related power 191–3

Japan
and LoS 47
and whaling 51–3, 193
and whaling boycotts 54
junkyard orbit 107
JUSSCANZ 195

Keohane, Robert O. 40, 196–7
and Nye, Joseph S. 192
knowledge
and climate change 143–4
consensual scientific and technical
205–7
function of regimes 38, 160
influence of on cognition 204–7
and ITU 117
of stratospheric ozone depletion 125
and whaling regime 55
Koh, Tommy 27
Kosmos 954, re-entry of 105
Krasner, Stephen, D.
regime definition 20
Structural Conflict 200
krill 75, 220
Ku band *see* geostationary orbit
Kyoto Protocol *see under* Framework
Convention on Climate Change

launchers, space 97–9
Law of the Sea (LoS) 45–8
Law of the Sea Convention (United
Nations Third, Montego Bay 1982)
47, 159, 179
EIF 47
and marine pollution 58
LDCs (less developed countries) 156
and Antarctica 81
and GSO 114, 116, 200
and stratospheric ozone 129, 131
and UNCLOS III 46
see also South *and* Group of 77
LEO (low earth orbit) 97, 105–6
liberal
institutionalism 172, 195–6, 201,
207–10, 213
internationalism 201
utilitarianism 195–9, 207–10
liberalism 201
Limits to Growth 135
London (dumping) Convention 1972
55–61
effectiveness of 177
'Long March' ELV 97
'low politics' 201

Madrid Protocol 1991 *see under* Antarctic
Treaty
Malaysia
and Antarctica 81
and forests 193
managerialism 217
manganese nodules 61
Marconi Company 99, 112, 188
Marine Mammals Protection Act
187
marine pollution
and epistemic communities 205
regime assessment 180, 197–8
regional arrangements 57
sources and extent of 55–6, 177
market failure 3, 5, 196
see also public goods
market principles 31–2, 196
MARPOL 1973 56–61, 159, 162, 167,
170, 197
effectiveness of 177
Marxism 185, 186–8
and regimes as epiphenomena 21

maximum sustainable yield (MSY) 50
measurable objectives 167–70
mercantilism, neo- 196
methane 135
microeconomic theory 40
minke whale 49
Molina-Sherwood hypothesis 125
Molniya orbit 97
monitoring *see under* compliance
Montreal Protocol 126–33, 159, 169,
    170–2, 176, 191, 201, 223–4
  cognitions 205–6
  compensation 129, 131–2
  control measures 129, 131, 148–9
  Copenhagen Meeting 127–30
  data reporting 132
  EIF 130
  equitable control 129
  London Meeting 127–30, 191, 192,
    194
  Meetings of the Parties 130
  as a paradigm 124, 146
  participation and LDCs 127, 129
Moon 99
  Treaty (1979) 100–1
Motorola Corporation, 96
Murdoch, Rupert *see* News Corporation

NASA
  quarantine precautions 105
  and remote sensing 109
  and space debris 107
navigation rights 45
New International Economic Order
    (NIEO) 31–2, 193–4, 221
  and deep seabed 64
New World Information and
    Communications Order (NWICO)
    32, 108–9, 118, 193
News Corporation 110
Nimbus 7 satellite 126
nitrous oxide 134
Non-Aligned Movement 193
  and Antarctica 88
non-governmental organizations
    (NGOs) 25, 27–9, 215, 224–5
  and compliance monitoring 168,
    174
  Environmental Liaison Centre
    International 28

  expertise 28
  political influence of 202
  UN observer status 28, 156–8
  and whaling 52
non-regimes 163–5, 179
norms 33–4
North Atlantic Marine Mammals
    Commission (NAMMCO) 52
North–South 195, 220, 221
  dialogue 32, 64
  resource transfers 176
Norway, whaling policy 51–2, 54,
    193

OECD 138, 191, 192, 221
oligarchy 190–1
Olson, Mancur 10
  see also collective action problem *and*
    public goods
orbit/spectrum resource 111–17, 197
  allocation and assignment 116–17
  1988 ITU plan 116–17, 194
  regime assessment 180
organizational strategies 193–5
Outer Space Treaty 8, 100, 115
ozone *see* stratospheric ozone layer
Ozone Fund, Interim 127, 195, 199
Ozone Trends Panel (UNEP) 126

Pardo, Arvid
  comments on seabed mining 62
  disappointment with LoS outcome
    47
  1967 UN Seabed speech 6
Pareto optimality 19
parallel system of seabed mining
    63–4
Part XI (deep seabed provisions of the
    Law of the Sea Convention 1982)
    47, 61–8, 179, 191, 194
participation, in regimes 156–8
precautionary principle
  and Antarctica 83
  and climate change 142
  and marine pollution 57–8
  as a norm 33
  and space debris 106
  and stratospheric ozone depletion
    128

principles
  allocative 30-3
  distinction from norms 30
  'prior consent' 111, 180, 197
prisoners' dilemma 10, 19
problem solving 155, 175-9
promotional regimes 163-5
public goods 5-6, 132, 189, 196
  defined as non-rivalness and non-
    excludability 4
  free riding 11, 174
public opinion
  cyclical 202
  ethical 201-4

radical political ecology 213, 222
radio communications 8
  see also frequencies
Radio Regulations (ITU) 117
radioactive
  fall-out 123
  material in orbit 105, 108
  waste and London Convention 61,
    201
Radiotelegraph Conferences 113
Reagan administration
  hostility to seabed regime 64, 203
  and ozone regime 202-3
  withdrawal of Moon Treaty 101
realism 154, 167, 172-3, 186, 199-200,
  207-9, 216
  neo-realist explanation 186-90
  and role of institutions 21
  see also hegemonic stability
Recognized Private Operating Agencies
  (RPOAs) 115
regime
  alternative meanings of 21-2
  change 42-5, 184, 212
  defined 17, 20-1
  strength typology 162-6
regime analysis
  assumptions of 21
  introduced 17
regime theory
  existence of denied 23
Regional Economic Integration
  Organization (REIO) 26
  see also European Union
remote sensing 109-11

res communis 4
  and Antarctica 80
  and GSO 114
  see also common heritage of mankind
res nullius 4, 17, 47, 223
  see also commons, open access
research communities 204-6
reservations 159
Rio Declaration 1992 34
  see also United Nations Conference on
    Environment and Development
    and sustainable development
Rio Summit 1992 *see* United Nations
  Conference on Environment and
  Development
rules 36-9
Russia
  decline as space power 104
  possible benefits of climate change 135
  and space debris 104-8

satellites
  uses of 95-9
  see also ASAT, DBS, orbit/spectrum
    resource *and* remote sensing
Saudi Arabia, ban on DBS dishes 111
scientific activity, in Antarctic 90-1
scientific consensus 28, 204-7
  and IPCC 143-4
  lack of in IWC 55
  in ozone and climate issue areas 134-5
seabed *see* deep seabed
seals, Antarctic 75
  Convention on Conservation of
    (CCAS) 77
self-management 171, 212
sinks 135, 223
Sky TV 110-11
Smith, Adam 195
'soft law' 159-60
South, the 14, 137
Soviet Union
  in Antarctica 86, 91
  and ASATs arms control 102-4
  disregard of whaling rules 54, 171
  Glavkosmos 107
  and liability for Kosmos 954 107
  and Montreal Protocol 132-3, 170
  naval power and LoS 45
  in space 97, 192

space, outer
  arms control 102–4
  commons 99–100
  debris 104–8
  delimitation of 100
  law 186
  vehicles 97–9
  war 95–6
  see also Outer Space Treaty *and*
    COPUOS
SPOT 104, 109
Sputnik 1 95
standard setting 37
STAR (Satellite TV Asia Region) 99,
  110
state
  -centric views 25, 199–200, 215,
    219
  role in commons management 15,
    25–6, 223
Stockholm 1972 *see* United Nations
  Conference on the Human
  Environment
straddling stocks (of fish) 49
Strange, Susan 152, 186
Strategic Defense Initiative (SDI) 101
stratospheric ozone layer 124–5
  depletion and consequences of 125
  and epistemic communities 205–6
  estimates of restoration 176
  regime assessment 180–1
  see also CFCs, Montreal Protocol,
    Vienna Conference and
    Convention
Strong, Maurice 26
structural
  explanations 184, 185–9, 206–9
  power 187
subsidiarity 213
sustainable development 1, 15, 32, 43,
  175

terra nullius 4
Thatcher, Margaret
  'conversion' to green issues 202
  hostility to seabed regime 64
Tolba, Mostapha Kamal 26
Toronto Group 125
total allowable catch (TAC) 53
trade and environment 218

tradeable permits 31
'tragedy of the commons' 10–15
  in fisheries 44
  in whaling 48–9
transaction costs 12, 196–7, 219
transborder data flows 102
transboundary air pollution 123–4
transnational corporations 27
transnational technical communities
  205–6, 220
'transparency' 161, 168, 174
Truman Proclamation 45, 187

UNESCO 108–9, 193
United Kingdom
  and LoS 47, 64, 186
  and UNESCO 194
  whaling activities and policy 50
United Nations 158, 160
  and Antarctica 92
  Disarmament Committee and space
    weapons 104
  General Assembly 47, 80, 88, 92, 122,
    137, 219
  Secretary General's consultations on
    LoS 47, 67
  system coordination 35
  voting system 160
United Nations Commission for
  Sustainable Development 216
United Nations Conference on
  Environment and Development 1992
  (UNCED) 1, 34, 57, 137–8, 156,
  202, 212, 215, 222
United Nations Conference on the
  Human Environment 1972
  (UNCHE) 1, 30
  and marine pollution 56, 61
  Principle 21 34, 221
  and whaling moratorium 54
United Nations Conference on the Law of
  the Sea (UNCLOS III) 7, 46–8, 179,
  189–90, 200, 218
  UNCLOS I and II 45
United Nations Environment Programme
  (UNEP) 26
  and FCCC 136–7
  Montreal guidelines 56
  and Montreal Protocol 126, 130
  Regional Seas Programme 57

United States of America
  and ASAT 102–4
  bureaucratic politics 202–3
  carbon dioxide emissions 135
  domestic legislation 187–8
  deep seabed policy 67
  and FCCC 137–41
  hegemonic position of 186–8, 203
  and Montreal Protocol 125, 126
  naval power and LoS 46, 199, 203
  and prior consent 111
  private sector and ITU 27
  in space 97
  and UNESCO 194
  unique pluralism 203
  veto role 209
  whaling activities and policy 50, 51, 54

value change 209
verification 167–70
veto role 191, 209
Vienna Conference and Convention
  (Ozone) 128, 158, 179
voting rules 160–1, 193

WARC (World Administrative Radio
  Conference) 118, 200
1977 broadcast 110

1979 general 114, 191
1992 general 115
1985–88 orbit 114, 116–19, 194
whaling
  Antarctic sanctuary 54, 77
  collapse of industry 49, 77
  commercial moratorium on 54–5
  International Convention for the
    Regulation of Whaling (ICRW)
    49–50, 158, 160, 165, 175
  International Whaling Commission
    (IWC) 51–2
  New and Revised Management
    Procedures 54–5
  regime assessment 180, 193, 198
  transformation of regime 201
  Washington Conference 1946 187
  watching 51
wilderness values 33, 77, 201
'win, win' solutions 39, 203
women, and cooperation 12, 19
Working Group I (of IPCC) 140
World Climate Conferences 136, 144
world government 16, 153, 222
World Meteorological Organization 136
world park, Antarctic as 77, 82, 90

Young, Oran, R. 40, 196
  institutional bargaining 208–9

9 780471 985747